SpringerBriefs in Geography

SpringerBriefs in Geography presents concise summaries of cutting-edge research and practical applications across the fields of physical, environmental and human geography. It publishes compact refereed monographs under the editorial supervision of an international advisory board with the aim to publish 8 to 12 weeks after acceptance. Volumes are compact, 50 to 125 pages, with a clear focus. The series covers a range of content from professional to academic such as: timely reports of state-of-the art analytical techniques, bridges between new research results, snapshots of hot and/or emerging topics, elaborated thesis, literature reviews, and in-depth case studies.

The scope of the series spans the entire field of geography, with a view to significantly advance research. The character of the series is international and multidisciplinary and will include research areas such as: GIS/cartography, remote sensing, geographical education, geospatial analysis, techniques and modeling, landscape/regional and urban planning, economic geography, housing and the built environment, and quantitative geography. Volumes in this series may analyze past, present and/or future trends, as well as their determinants and consequences. Both solicited and unsolicited manuscripts are considered for publication in this series.

SpringerBriefs in Geography will be of interest to a wide range of individuals with interests in physical, environmental and human geography as well as for researchers from allied disciplines.

Bin Li

Resilient Governance of Urban Redevelopment

State, Market and Communities in China Since 1990

Springer

Bin Li
Faculty of Innovation and Design
City University of Macau
Macau, China

Peking University Shenzhen Graduate
School
Shenzhen, China

ISSN 2211-4165   ISSN 2211-4173 (electronic)
SpringerBriefs in Geography
ISBN 978-981-99-2927-6   ISBN 978-981-99-2928-3 (eBook)
https://doi.org/10.1007/978-981-99-2928-3

© The Editor(s) (if applicable) and The Author(s) 2023. This book is an open access publication.

Open Access This book is licensed under the terms of the Creative Commons Attribution 4.0 International License (http://creativecommons.org/licenses/by/4.0/), which permits use, sharing, adaptation, distribution and reproduction in any medium or format, as long as you give appropriate credit to the original author(s) and the source, provide a link to the Creative Commons license and indicate if changes were made.
The images or other third party material in this book are included in the book's Creative Commons license, unless indicated otherwise in a credit line to the material. If material is not included in the book's Creative Commons license and your intended use is not permitted by statutory regulation or exceeds the permitted use, you will need to obtain permission directly from the copyright holder.
The use of general descriptive names, registered names, trademarks, service marks, etc. in this publication does not imply, even in the absence of a specific statement, that such names are exempt from the relevant protective laws and regulations and therefore free for general use.
The publisher, the authors, and the editors are safe to assume that the advice and information in this book are believed to be true and accurate at the date of publication. Neither the publisher nor the authors or the editors give a warranty, expressed or implied, with respect to the material contained herein or for any errors or omissions that may have been made. The publisher remains neutral with regard to jurisdictional claims in published maps and institutional affiliations.

This Springer imprint is published by the registered company Springer Nature Singapore Pte Ltd.
The registered company address is: 152 Beach Road, #21-01/04 Gateway East, Singapore 189721, Singapore

Preface

Urban redevelopment in Chinese cities is significant for urban growth. Through focusing on Guangzhou, the third largest city in China and a pioneer in post-1978 reform, as the representative, this study aims to investigate the dynamics of governance of urban redevelopment in Guangzhou, China from 1990 to 2015 under the Chinese authoritarian regime. The connections between such governance and the authoritarian regime are the key to understand the governance dynamics; this is the main contribution of this research.

Through analyzing data collected from semi-structured interviews, secondary data and participant observation, this research reveals that, firstly, in an authoritarian-style institutional framework, a land-oriented growth and redevelopment context has been formulated; second, there are three phases with different patterns of governance: (1) the Primitive Market Phase (1990–1998); (2) the Pure Government Phase (1998–2006); and (3) the Multiple Players Phase (2006–2015); third, the purpose of changes in governance mode between different phases is displayed as pursuing economic growth in various developmental environments, governance modes are adaptive and resilient to respond to changed internal and external circumstances, Chinese authoritarian regime is the key to support such adaptive capacities in governing urban redevelopment in Guangzhou resiliently.

Macau, China   Bin Li

Acknowledgments

This research and publication are funded by National Natural Science Foundation of China (41871154), National Natural Science Foundation of China and Science and Technology Development Fund of Macau (52061160366), (0039/2020/AFJ), Peking University Lincoln Institute Research Funding 2018–2019 and 2021 Tiehan Open Research Funding from Peking University Future City Lab (Shenzhen).

Foremost, I am grateful to my parents, Mr. Tieming Li and Ms. Guifang Deng for without their unconditional love, support, and grace, my study and my life would not have been possible.

I would like to thank the help and support of my supervisors, Dr. Lauren Andres, Dr. Vlad Mykhnenko, and Dr. Phil. Jones. Thank you for their constant guidance and encouragement, and pushing me to follow the schedule when I always want to delay my progress. I would also like to thank Dr. Oleg Golubchikov who worked as the leading supervisor for almost 1 year and helped me a lot.

I wish to acknowledge the support and company received from Li Manning; she always gave me energy to continue my study. I also want to acknowledge Dr. Mao Feng for his academic insights and help in improving my thesis. Dr. Gong Zhaoya, Dr. He Shaoying, Huang Lihua, Dr. Zhu Chenwei, Luo Qiliang, Xiao Yazhi, and Zheng Yue also contribute to my project. Another thanks should be sent to my Ph.D. colleagues in room 225 over the years, which made my stay and studies at GEES more enjoyable. In particular, thanks to Arshad, Bobby, Colin, Deyala, Elly, Janna, Kiseong, Phil, Taesuk, Tess, Thom, Tom, Yixin, Yiting, and Upuli. Thank you all for your friendship. I also would like to thank profusely all staff of GEES (especially Gretchel), the Graduate School and ISAS for their kind help and cooperation throughout my study period.

I am really thankful to Doris, Dr. Li Suinong, Dr. Liu Lixiong, Dr. Liu Yao, Dr. Wei Lihua, Li Xinheng, Prof. Fang Yuanping, Prof. Li Xun, Prof. Li Zhigang, Prof. Wang Shifu, Prof. Yuan Qifeng, Wang Min, Wang Min, Xian Yueshi, Xing Xiaowen, Zhang Longlong, Zeng Chunxia, and Zhou Kebin, who have helped me a lot in two field trips in Guangzhou, China. I also want to show my thanks to A Li, Deng Bingtao, Dr. Deng Kanqiang, Dr. Feng Jiang, Dr. Huang Huiming,

Dr. Ye Haojun, He Shan, Huang Qingkan, Huang Wei, Liu Tianhe, Liu Xuan, Prof. Huang Dongya, Prof. Liu Yungang, Prof. Wu Zhigang, Xu Jiaquan, and Xu Shiguang, they also provided helpful information to my research.

Contents

1	**Introduction**	1
	1.1 Miraculous or Miserable? Urban Redevelopment in Economic Growth of China	1
	1.2 Literature Review and Gaps	3
	1.3 Research Design	6
	1.3.1 Aims and Objectives	6
	1.3.2 The Reasons to Choose Guangzhou After 1990	6
	1.3.3 Structure of This Study	8
	References	10
2	**Background of Urban Redevelopment**	13
	2.1 Introduction of Guangzhou	13
	2.2 China's Authoritarian Political Regime	14
	2.3 Relation Between Guangzhou and the Central Government	15
	2.3.1 Central-Local Relations and 1994 Reform	16
	2.3.2 Leaders of Guangzhou in a Promotion Tournament	18
	2.4 A Land-Oriented Growth Structure and the Role of Urban Redevelopment	20
	2.4.1 Land Property Rights and Regulations	21
	2.5 An Institutional Framework of Land-Oriented Growth and Redevelopment	25
	References	27
3	**Three Phases of Urban Redevelopment Governance**	31
	3.1 Primitive Market Phase (1990–1996)	31
	3.1.1 Overall Characteristics of Redevelopment in This Phase	32
	3.1.2 Typical Project: Liwan Square Redevelopment (1992–1995)	35
	3.1.3 Summary of the Primitive Market Phase	38
	3.2 Pure Government Phase (1998–2006)	38
	3.2.1 Overall Characteristics of Redevelopment in This Phase	39
	3.2.2 Typical Project: Three Changes of Guangzhou	39

		3.2.3 Summary of the Pure Government Phase	41
	3.3	Multiple Players Phase (2006–2015)	42
		3.3.1 Overall Characteristics of Redevelopment in This Phase	42
		3.3.2 Typical Project: Enning Road Redevelopment (2006–2015)	43
		3.3.3 Summary of the Multiple Player Phase	46
	3.4	Guangzhou Style of State-Market-Community Relations	46
	References		47
4	**Resilient Governance of Urban Redevelopment**		51
	4.1	Changes Impacted on Urban Redevelopment in Guangzhou	51
		4.1.1 Political Changes	51
		4.1.2 Social Changes	55
		4.1.3 Economic Changes	58
	4.2	Resilient Governance as Responses to Changes	59
	4.3	Authoritarian Characteristics as the Basis for Resilient Governance	61
		4.3.1 Autonomies of Guangzhou Municipality	61
		4.3.2 Mayors' Preference to Influence Urban Redevelopment Governance	62
		4.3.3 Institutional Capacities to Support Resilient Governance	65
	References		66
5	**Conclusion**		69
	5.1	Research Findings	69
		5.1.1 Institutional Background	69
		5.1.2 Three Phases of Governance in Urban Redevelopment	70
		5.1.3 Resilient Governance of Urban Redevelopment in Guangzhou	71
	5.2	Comparing Guangzhou to Other Chinese Cities	72
	5.3	Limits of This Research and Areas for Future Research	75
	Reference		76

Chapter 1
Introduction

Abstract This research focuses on the political nature of urban redevelopment from a governance perspective; characteristics of its governance will be analysed within the Chinese authoritarian regime. Urban redevelopment is chosen as the research object because it is significant and complex enough to be a typical case in Chinese economic growth. The main problem in urban redevelopment is the issue of social cooperation among different stakeholders; it is the problem of governance. Theories of governance origin from the developed world and their applications into Chinese context will bring about interesting development of governance studies and exciting reflection of Chinese politics. In this perspective, the most important characteristic of China's political system, an authoritarian regime can strongly affect the characteristics of governance. Guangzhou, as the third largest city in China and the political-economic centre of south China, has been chosen as the studied site to reveal its political nature in Urban redevelopment. Based on this analysis, a resilient mode of governance in Guangzhou redevelopment can be established. This mode is a small fragment of mirror to reflect the whole political picture of China.

Keywords Urban redevelopment · Resilient governance · Authoritarian regime · China · Guangzhou

1.1 Miraculous or Miserable? Urban Redevelopment in Economic Growth of China

Urban redevelopment is the object of this study because redeveloping Chinese metropolises is an important and controversial element in the narrative of China's economic growth. This growth in the last few decades is significant. In terms of the ranking of GDP (gross domestic product) worldwide, China has upgraded from the eleventh in 1990 to the second in 2010. As the result of this rapid productivity growth, living standards and personal income have also significantly improved in this period (Lin 2011). Urbanisation is an important part of such economic expansion. Bai et al. (2012) claim that urbanisation and economic growth are closely connected, according to the analysis of relation between the expansion of built-up areas and the growth of GDP per capita in China. Sun (2011) indicates that GDP per capita can increase by 4.34% when the degree of urbanisation (indicated by the percentage of urban population in the whole population) has increased by 1%.

In terms of urbanisation, the urban population in Mainland China has increased from 18 per cent in 1978 to 52% in 2012; the urban population has grown from 352 to 670 million from 1995 to 2010 (Ye and Wu 2014). Millions of migrants have moved into cities, especially metropolises. The pursuit of economic growth is a huge driving force for the expansion of construction land and upgrading of building environments in urban space. Such changes led to a more efficient and competitive urban economy emerging in China after the mid-1990s (Lin 2011). Indeed, Stiglitz, one of the winners of the 2000 Nobel Memorial Prize in Economic Sciences, states that Chinese urbanisation and high-tech innovations in the United States are the two keys for the development of the whole world in the twenty-first century (People's Daily Online 2005).

Urban redevelopment is an influential activity in extra-large Chinese cities as a part of their urbanisation processes; for the involved population such redevelopment might be miraculous or miserable at the same time. Urban redevelopment has been powerfully driven by Chinese urbanisation while huge amounts of immigrants have stimulated demands for space in urban areas. Redevelopment is more significant in extra-large cities, such as Beijing, Shanghai, Guangzhou and Shenzhen, which have populations of more than 10 million (People's Daily Online 2005). This is because these metropolises are the most active points for economic growth and migration. However, development in these extra-large cities has been controlled in terms of the availability of new construction land according to national policy; redevelopment in existed construction land is a better choice than developing newly expanded construction land under the political regime (He et al. 2006; Lin 2011).

Urban redevelopment is exciting, dramatic and significant in terms of building brand new departments, shopping malls and offices; but it also brings social conflicts, destroying historical buildings and obliterating Chinese culture. Campanella (2010) in his book, The Concrete Dragon: China's Urban Revolution and what is meant for the World, devotes one chapter, Chap. 5 City of Chai (demolish), to describe urban redevelopment in China. He quotes the work from Huang Rui, a Chinese artist, named Chai-na/China as a symbol of urban redevelopment. Chai (拆) means demolishing and destroying in Chinese; its spelling and pronunciation seem to be similar with the word 'China'. In Huang's (2005) view, demolition and destruction might be the most important phenomena in China.

In mass media of China, urban redevelopment is a hot topic. In 2007, a demolition project in Chongqing, a large city in South West China, became the focus of mass media (Tian 2007). The resident of a single house refused to be removed from her home and the house remained while every house around it was demolished; therefore the one remaining house became an island in which the homeowner still lived within an ocean of demolition (Zhou 2007). This dramatic image shocked the whole country; problems in redevelopment entered the field of public debate.

Two years later, on the 13th November 2009 in Chengdu, another large city in South West China, a lady named Fuzhen Tang burned herself to death on the roof of her house to protest against the violent demolition of her property (Sina News 2009). This stimulated more public discussion about redevelopment in television, newspapers and the internet. It seemed that public consensus about protecting property rights, the legal process of governmental behaviour and definition of public interest in redevelopment had been awakened through the stimulation of public debates.

In the last few decades urban redevelopment in China, as a consequence of the urbanisation process in China's economic growth, is significant in terms of its achievements and conflicts. This process might be described as both miraculous and miserable. It seems the beginning of Dickens' A Tale of Two Cities has never been out of date. For its significance and complexity the nature of urban redevelopment deserves to be comprehensively and systematically analysed. Numerous literatures have already explored this topic; my research will join this group of studies from a specific perspective.

This research investigates urban redevelopment from the theoretical perspective of governance and analyses this governance in an authoritarian environment. Governance is a popular concept to analyse urban redevelopment in Chinese metropolises because ideas around governance can provide a deep understanding of the political nature of society. Under the authoritarian system, crucial challenges have been reflected in governance level, namely, how to bring together the various forces that drive development into harmonious relationships and how to solve conflicts.

Redevelopment plays such a crucial and controversial role in China's development; it is an interesting object for the study of governance. Another reason to support this study is that urban redevelopment is a special objective as an institutional function under authoritarianism. Every important law and rule, such as property rights, tax and fiscal income, legal process, the planning system and autonomy of communities, is reflected in urban redevelopment. Urban redevelopment is a mirror of the whole Chinese political regime; it is also a mirror of China's society.

1.2 Literature Review and Gaps

Urban redevelopment, accelerated from the late 1980s, is an important part in urban development and transformation, because stricter regulation of land conversion in suburban space has brought urban development back to the inner cities (He et al. 2006). Briefly, research about urban redevelopment and its governance in China have three main concerns, institutional background, property rights of land, and governance in this field. Redevelopment in Chinese cities is based on specific institutional arrangements. Housing reform has led to its marketisation, which in turn has brought prosperity to the urban housing market. Redistribution of power between the central and local state has moved control of property rights over urban land into the hands of local authorities. The local state build-up of the land leasing system through auctions could transform potential land values in the property market into revenue for local authorities (Wu 2002; He and Wu 2005, 2009; He 2007).

Property rights of land in Chinese cities have been defined as land use right in urban land; the local states are the agencies to actually control use rights of land in cities when these lands are nominally belonging to the whole population of China. Such use rights of land have been described as ambiguous; usage rights and developmental rights are separated into the hands of different stakeholders, such as work units, residences, planning and land management departments. The socialist use rights of property are described as economic rights (de facto rights) rather than legal rights; these de facto rights in public housing, work-unit housing and village housing are unsecured and uncertain for long-term consideration. Therefore, potential values of land are open to competition from different agents in the field of urban redevelopment. Capacities to compete are relevant to the hierarchical position of agencies (Zhu 2002, 2004; Lin and Ho 2005; Tian 2008).

Based on this institutional background and property rights, land owners, the local state and developers have formed a coalition to accelerate redevelopment by assembling fragmented property rights. The state has strong incentives to benefit the interests of developers in terms of financial and administrative support, beautification construction and infrastructure improvement; because the local state aims to grasp revenue through promoting the urban image and attracting investment. These activities of the state can also be understood as an extension of the state to newly-emerged field of redevelopment with economic opportunities. Social interests are sacrificed to support such redevelopment in terms of excluding public participation. This mode of governance has some neoliberal characteristics in terms of the relations between market and state, and the approach of the local state to benefit market players.

This reform of governance is responsive to crisis and difficulties rather than a designed blue print. Such governance includes different and paradoxical logics, such as economic growth, consensus among people and social stability. These logics lead to various governance modes, from the cooperation mode to self governance mode, to adapt to increased unstructured complexity (Zhang 2002; Wu 2002; He 2007, 2012; He and Wu 2005, 2009; Lin et al. 2015).

Such governance patterns have been changed in different periods of redevelopment. He (2012) describes two waves of redevelopment and its governance in Guangzhou after the late 1980s. between these two waves, the scale of redevelopment has been transformed from small to large scale; the purpose of projects has changed from improving living condition to reimaging a global city; the result of redevelopment has been changed from increasing use value to exchange values. She concludes that these changes act towards a Neoliberal governance mode in terms of optimal strategy for capital accumulation, such as reducing uncertainty of projects, intentional land deficiency and assembly of fragmented land ownership.

Interaction between entities from state, market and communities in Chinese cities have also been analysed through the regime approach. Zhu (1999) analyses coalitions between local authority, SOEs (State-Owned Enterprises) and developers to realise both economic growth and social stability. He finds that the local state created an attractive climate for industry to increase the tax base and revenue, SOEs contributed to social stability in terms of offering jobs and welfare for workers and developers brought capital and experience into redevelopment. Zhang (2002) elaborates the

1.2 Literature Review and Gaps

socialist regime characteristics in Shanghai, citing 'strong local government followed by cooperative nongovernmental sectors with excluded community organization' (Zhang 2002, p. 475).

From a regime perspective, this socialist regime is different from its original model in the US because the private sector is relatively less influential and communities in China is still weak. Yang and Chang (2007) also focus on Shanghai to reveal a public–private partnership as a pro-growth regime to developed Zhang's study in three aspects: isolating the district government as a player with autonomy from the municipality, analysing rent-seeking opportunities in the redevelopment process and pointing to the lessened impacts of the central government. Li and Li (2011) argue that an urban coalition has emerged between public and private actors in the redevelopment of urban villages in Shenzhen, through the biased distribution of resources and urban planning policy.

Since 2009, a new group of policy, the 'Three Old Redevelopment Policy', has been experimented in Guangzhou, Foshan and Dongguan in Guangdong Province; an interests-sharing coalition has been established by this policy to include government, developers and communities in redeveloping processes (Li and Liu 2018). Renters, as property users rather than owners, have been excluded from such a coalition, even in upgrading projects as urban regeneration (Li et al. 2021).

In these studies about urban governance in redevelopment in Chinese cities, the main difference from western cases is China's authoritarian regime which consistently is the dominant factor in urban politics. This authoritarian defines institutional background, property rights of land and governance with Chinese characteristics. In terms of institutional arrangement, the local states are pushed by the central authority to produce land-based revenue; in property rights, use right is a convenient tool for the local state to steer and intervene market behaviours; in governance, the local is the dominant role to establish redevelopment strategies to benefit some selected actors by the supports of public funding and administrative resources.

Besides, after Chinese authoritarian regime has organised quite different pathway of urban redevelopment and economic growth, Chinese cities and western cities have both displayed similar Neo-liberal patterns of governance. In these similar characteristics, the state mobilise public resources to advance the interests of private capital while the public interest might be sacrificed (He and Wu 2009).

Literatures about urban regimes in redeveloping Chinese cities also revealed the influence of China's authoritarian regime in terms of the role of the state in regimes. Chinese authoritarian style state is more powerful in growth-coalitions in Chinese cities when the private sectors are more influential in American cases. However, such different function of the state in Chinese cities and American ones has produced similar results. The state and developers are included in growth-coalition and communities are more or less exclude in such activities.

The biggest gap in these literature is the lack of real political concern on urban governance; which means that the most important political background of urban governance in China, the authoritarian regime, has not been analysed clearly. My research will connect these two ideas, the authoritarian regime and urban governance, to investigate and establish the special mechanism in Chinese cities. Namely, how could the authoritarian regime affect governance modes and their changes, and how could the governance modes contribute to the survival and development of the authoritarian regime. In reality, the authoritarian regime and urban governance are contemporaneous; however, their separation in a theoretical perspective is meaningful because it emphasises the most important Chinese factors.

1.3 Research Design

1.3.1 Aims and Objectives

This study aims to investigate the political nature of urban redevelopment in Guangzhou in terms of connecting two main aspects, the governance of redevelopment and characteristics of an authoritarian regime. The ideas about governance might be very interesting to be applied in analysing politics in Chinese cities because they originate from a Western context, especially from the UK and some phenomena in governance that have been studied in the UK also appear in Chinese cities.

In the mirror of urban redevelopment, the whole picture of the Chinese political landscape has been reflected. Therefore, the aim of my research is to reveal the political nature of urban redevelopment in an authoritarian environment. The three main research objects of this study are focused on redevelopment, governance and the authoritarian regime; such three objects are interacted in my study. Redevelopment is the raw material for research; features of governance are abstracted from social interaction in redevelopment; dynamics of governance are affected by Chinese authoritarian regime. Research objectives are established from interaction between three objects, namely, the characteristics of redevelopment, governance and the authoritarian regime, and the dynamics of the relations between governance and Chinese authoritarian regime.

In order to achieve a deep understanding of the connection between the governance of urban development and the prevailing authoritarian regime I will choose Guangzhou, a pioneer of China's reform, as the case to study this topic. The analysis of urban redevelopment in Guangzhou from 1990 to 2015 is the empirical object which facilitates investigation of the political system in Chinese cities.

1.3.2 The Reasons to Choose Guangzhou After 1990

My research aims are concerning about political nature of urban redevelopment in Chinese cities under an authoritarian regime. From this perspective, which Chinese city is appropriate to be analysed? Extra-large Chinese cities is the hot pot of urban redevelopment. In terms of extra-large cities, Beijing, Shanghai, Guangzhou and Shenzhen are considered as the 'First-tier Cities', a very popular idea in Chinese

1.3 Research Design

population, among Chinese cities because of their significance in political, economic and cultural system of China. Therefore, any of these four cities are deserved to be analysed as representative of Chinese cities in terms of their importance. Especially these cities are national economic centres; which means their urban redevelopments are more attractive to capital to invest for profits. It results in active patterns of governance in urban redevelopment. In these four cities I choose Guangzhou, the capital of Guangdong Province, a national central city in south China, as the studied case for two reasons.

First, Guangzhou is a pioneer city in China's reform; therefore, narratives in this pioneer can reveal some dynamics which cannot be observed in other cities. This is because other Chinese cities as followers in China's reform often learn experience from Guangzhou and Guangdong; these followers are not original place for policy experiments. These experiments can better reveal political nature of urban issues because they are usually the frontline of reform with fierce conflicts between different social groups. Guangdong Province is a traditional pioneer in China's transition while it is far away from the political centre, Beijing, and near the most Westernised Chinese city, a global metropolis and ex-colony city, Hong Kong. Therefore it is safer to do some political-economic experiments in Guangdong Province than in Beijing and Shanghai which are more crucial for the Chinese Communist Party (CCP) to control the whole country.

What happens in Guangdong Province often acts as an experiment and provides good experiences to other region of China to imitate. Shenzhen, one of the most important cities in Guangdong, is the most pioneering city in China's transition. However Shenzhen is a special case because it is a special experimental zone in China's reform. It has just been developed after the beginning of transition in 1978. It has only thirty-eight years history as a city until now; which means it is less typical with traditional characteristics as Chinese cities, such as powerful local bureaucratic group, long-term urban culture and historical heritages. Therefore, Guangzhou, the capital of Guangdong Province with more than 2000 years history, is a good case for researching patterns of governance in Chinese cities.

At the same time, because Guangzhou is closer to Hong Kong; it has been influenced to be a more democratically-governed city compared to Beijing. It brings about more meaning to Guangzhou as a pioneer in transition towards to more open, transparent and democratic system. It also may reduce the difficulty of investigation in Guangzhou than in other large Chinese cities because of the relative openness of government (Lu 2001; Zhu and Zhang 2004).

Second, Guangzhou has been selected as one of the first cities to experiment the 'Three old redevelopment' policy which has been launched in 2009. 'Three old redevelopment' includes the redevelopment of old villages, old cities, and old factories. This policy is coming from the Ministry of Land and Resources. Guangdong Province is the first place to experiment this policy. It brings about a series of special institutional arrangements to encourage urban redevelopment through releasing planning regulation, reducing taxation, providing financial resources and supporting to build partnership. Such institutional arrangements indicate some new tendencies in urban governance which are difficult to be found in other Chinese cities.

Foshan, Dongguan and Guangzhou are the experimental cities in this policy. Foshan is the first city among these three cities to apply this policy; however, it is a much less important city compared with Guangzhou in a national level. Besides, Foshan and Dongguan are not as important as Guangzhou in China. Therefore, Guangzhou has a unique position to be analysed in the experimental policy, 'Three old redevelopment' policy in my study.

My research will more focus on the years after 1990. This starting point is selected for three reasons.

After the Tiananmen Square Protest in 1989, the Chinese Communist Party (CCP) and its regime has paid much more attention to economic growth to survive the legitimacy crisis. Urban development and redevelopment is an important part of economic growth (Gries and Rosen 2004). This period has displayed stronger incentives of authoritarian regime to pursue land-based economic growth. Such incentives can reveal the significance of authoritarian factors in governing urban redevelopment as one of the research objectives.

After China's reform from 1978, urban lands were freely distributed to land users until 1987. From 1987 to 1992, a new institution of a market-oriented land using system has been established. After that, local authorities have more incentive to develop and redevelop urban land for fiscal income (Zhu 2005). This institutional arrangement formed the structure of urban redevelopment in Guangzhou and simulated the fast growth of redevelopment. In this way, it builds up a specific pattern of governance. It concerns another aspect of research objectives, the governance patterns.

After 1990, the growth of both Total Investment in Building and Residential Building Areas has been accelerated in Guangzhou. These two indexes displays increased investment in construction industry; and this industry is closely related to urban redevelopment. At the same period, the investment of the real estate in Guangzhou also boomed (see Table 1.1 and Fig. 1.1).

1.3.3 Structure of This Study

This research will be displayed as three parts in following chapters. Based on introducing basic information about Guangzhou, Chap. 2 aims to put urban redevelopment of Guangzhou into the institutional matrix within Chinese authoritarian regime. Incentive and resources of the local state and other actors to push redevelopment forwards will be figured out. Chapter 3 concerns about governance modes of urban redevelopment in Guangzhou after 1990 in terms of its whole picture and individuals cases. Three distinct phases will be defined according to different relationship between state, market and communities in redevelopment. Chapter 4 is a connection between authoritarian institutions (Chap. 2) and governance of urban redevelopment (Chap. 3) to understand the reasons and mechanism behind these changes of governance modes. After such a connection, a resilient governance mode would emerge to reflect the characteristics of China's political system.

1.3 Research Design

Table 1.1 Total investment and built areas in construction in Guangzhou

Year	1980	1985	1990	1995	2000	2005	2010	2013
Cross output value of construction industry (10,000 yuan)	46,450	153,486	355,333	1,816,313	2,561,326	6,331,382	12,805,28	21,828,895
Floor space of buildings competed (10,000 m^2)	159.95	285.22	439.08	890.81	1,150.36	1,598.79	1,509.20	2,556.74

Source Statistics Bureau of Guangzhou Municipality and Guangzhou Survey Office of National Bureau of Statistics (2014), p. 334

Fig. 1.1 Total investment of real estate in Guangzhou. *Source* Author's drawing based on data from Statistics Bureau of Guangzhou Municipality and Guangzhou Survey Office of National Bureau of Statistics (2014), p. 110

References

Bai X, Chen J, Shi P (2012) Landscape urbanization and economic growth in China: positive feedbacks and sustainability dilemmas. Environ Sci Technol 46(1):132–139

Campanella T (2010) The concrete dragon: China's urban revolution and what is mean for the world. Princeton Architectural Press, New York

He SJ (2007) State-sponsored gentrification under market transition, the case of Shanghai. Urban Stud 43(2):171–198

He SJ (2012) two waves of gentrification and emerging rights issues in Guangzhou, China. Environ Plann A 44(12):2817–2833

He SJ, Wu F (2005) Property-led redevelopment in post-reform China: a case study of Xintiandi redevelopment project in Shanghai. J Urban Aff 27(1):1–23

He SJ, Wu F (2009) China's emerging neoliberal urbanism: perspectives from urban redevelopment. Antipode 41(2):282–304

He SJ, Li ZG, Wu FL (2006) Transformation of the Chinese City, 1995–2005 geographical perspectives and geographers' contributions. China Information 20(3):429–456

Huang R 黄锐 (2005). Chai-na/China 拆那 [Demolition where/China]. [online] Available at: http://blog.ocula.com/post/119638385812/thedancingtoast-黄锐-huang-rui-chai-nachina. Accessed 22 Oct 2016

Kong WM 孔伍梅 (2008) Guangzhou jiuchengjuzhuqugaizao moshiyanjiu 广州旧城居住区改造模式研究 [The study on Guangzhou inner city: residential renewal]. Master Thesis, Sun-yat Sen University

Li B, Liu CQ (2018) Emerging selective regimes in a fragmented authoritarian environment: the 'Three Old Redevelopment Policy' in Guangzhou, China from 2009 to 2014. Urban Stud 55(7):1400–1419

Li B, Tong D, Wu YY, Li GC (2021) Government-backed 'laundering of the grey' in upgrading urban village properties: Ningmeng Apartment Project in Shuiwei Village, Shenzhen, China. Progress Plann 1–16

Li L, Li X (2011) Redevelopment of urban villages in Shenzhen, China–an analysis of power relations and urban coalitions. Habitat Int 35(3):426–434

References

Lin GCS (2011) Urban China in transformation: hybrid economy, juxtaposed space, and new testing ground for geographical enquiries. Chin Geogra Sci 21(1):1–16

Lin GCS, Ho SPS (2005) The state, land system, and land development processes in contemporary China. Ann Assoc Am Geogr 95(2):411–436

Lin Y, Hao P, Geertman S (2015) A conceptual framework on modes of governance for the regeneration of Chinese 'villages in the city.' Urban Stud 52(10):1774–1790

Lu D 卢狄 (2001) Guangdong gaigekaifangde sangefazhanjieduan jizhuyaochengjiu 广东改革开放的三个发展阶段及主要成就 [Three development stages & main achievements for guangdong since reform and open policy]. Special Zone Econ 特区经济 2001(7):5

People's Daily Online (2005) China to push forward urbanization steadily. [online] Available at: http://en.people.cn/200505/12/eng20050512_184776.html. Accessed: 22 Oct 2016

Sina News (2009) Chengdu chaiqianhu zifenshijian 成都拆迁户自焚事件 [Incident of a Chengdu Relocatee's self-burning]. [online] Available at: http://news.sina.com.cn/z/cdcqhzf/. Accessed: 22 Oct 2016

Statistics Bureau of Guangzhou Municipality and Guangzhou Survey Office of National Bureau of Statistics (2014) 2014guangzhou tongjinianjian 2014广州统计年鉴 [Guangzhou statistical yearbook, 2014]. China Statistic Press, Beijing

Sun WK 孙文凯 (2011) Chengshihua yujingjizengz guanxifenxi 城市化与经济增长关系分析—兼评中国特色 [The relationship between urbanization and economic growth: with a focus on China's distinctiveness]. Jingjililunyuguanli 经济理论与经济管理 4:33–40

Tian L (2008) The Chengzhongcun land market in China: boon or bane?—A perspective on property rights. Int J Urban Reg Res 32(2):282–304

Tian GF 田乾峰 (2007) Chongqing Zuiniu Dingzihu Xiang Shigaoyuan Tiqi Shenshu 重庆最牛钉子户向市高院提起申诉 [The most powerful Dingzihu sues in the high court of Chongqing] [online] [online] Available at: http://news.sina.com.cn/c/2007-03-24/014312598293.shtml. Accessed: 12 June 2017

Wu F (2002) China's changing urban governance in the transition towards a more market-oriented economy. Urban Stud 39(7):1071–1093

Yang Y, Chang C (2007) An urban regeneration regime in China: a case study of urban redevelopment in Shanghai's Taipingqiao area. Urban Stud 44(9):1809–1826

Ye L, Wu M (2014) Urbanization, land development, and land financing: evidence from Chinese cities. J Urban Aff 36(51):1–15

Zhang T (2002) Urban development and a socialist pro-growth coalition in Shanghai. Urban Aff Rev 37(4):475–499

Zhou M 周民 (2007) Chongqing Zuiniu Dingzihu Xiang Shigaoyuan Tiqi Shenshu 重庆最牛钉子户向市高院提起申诉 [The most powerful Dingzihu sues in the high court of Chongqing] [online] Available at: http://news.sina.com.cn/c/2007-03-24/014312598293.shtml. Accessed: 12 June 2017

Zhu J (1999) Local growth coalition: the context and implications of China's gradualist urban land reforms. Int J Urban Reg Res 23(2):534–548

Zhu J (2002) Urban development under ambiguous property rights: a case of China's transition economy. Int J Urban Reg Res 26(1):41–57

Zhu J (2004) From land use right to land development right: institutional change in China's urban development. Urban Stud 41(7):1249–1267

Zhu J (2005) A transitional institution for the emerging land market in urban China. Urban Stud 42(8):1369–1390

Zhu WH 朱文晖, Zhang YB 张玉斌 (2004) Gaigekaifangyilai zhongguoquyuzhengce desicitiaozheng jiqiyanpan' 改革开放以来中国区域政策的四次调整及其研判 [Four times revising of regional policy in Chinese reform]. Kaifangdaobao 开放导报 1:6

Open Access This chapter is licensed under the terms of the Creative Commons Attribution 4.0 International License (http://creativecommons.org/licenses/by/4.0/), which permits use, sharing, adaptation, distribution and reproduction in any medium or format, as long as you give appropriate credit to the original author(s) and the source, provide a link to the Creative Commons license and indicate if changes were made.

The images or other third party material in this chapter are included in the chapter's Creative Commons license, unless indicated otherwise in a credit line to the material. If material is not included in the chapter's Creative Commons license and your intended use is not permitted by statutory regulation or exceeds the permitted use, you will need to obtain permission directly from the copyright holder.

Chapter 2
Background of Urban Redevelopment

Abstract This chapter aims to analyse the geographic and institutional background, within an authoritarian regime, to understand the changing process of governance over the field of urban redevelopment. A brief introduction of Guangzhou will be displayed. Based on geographic information, relevant institutions are introduced. Institutional resources and power are distributed between different departments in government, and between government, markets, and communities. This distribution constrains and constructs the patterns of collective behaviours in redevelopment process in terms of influencing preference, locating resources and formulating strategies of different entities. Urban redevelopment is a crucial channel for authoritarian local state to pursue economic growth and political performance.

Keywords Urban redevelopment · Authoritarian regime · Guangzhou · Institutions

2.1 Introduction of Guangzhou

Guangzhou, with the nicknames of the Flower City and the Goat City, is the third biggest metropolis in China, and the capital of Guangdong Province. In 2013, Guangzhou produced a Gross Domestic Product (GDP) of 1,542,014 million China Yuan (CNY) (exchange rate between CNY and GBP is 1:9.64 at 20 May 2015), which is equal to 119,695 CNY per capita GDP (Statistics Bureau of Guangzhou Municipality and Guangzhou Survey Office of National Bureau of Statistics 2014). It is the political, economic and cultural centre of Guangdong Province and South China. As a sub-provincial municipality, it has more autonomy than other normal municipal cities. Guangzhou is located north of the Pearl River Delta around 100 kms away from both Hong Kong and Macau, two international cities and former colonies.

The city of Guangzhou was originally built more than 2,000 years ago and became the capital of the Nanyue Kingdom in 203 BC. The city was first named Panyu and was later named Guangzhou in AD 264 in the Han Dynasty. In 1757 Guangzhou was the only port still open to international businesses when other ports were forbidden to do so according to the Hai Jin (海禁) policy from the Qing Dynasty. It was a prosperous period for Guangzhou until the Opium War of 1840. In 1921, the Guangzhou municipal government was built as the first municipality in China in a modern sense (The editor committee of the chronicle of Guangzhou 1998).

Guangzhou was launched as an open city in the second batch of cities open to global business in 1984, after the first batch of four open cities (Shenzhen, Zhuhai, Shantou and Xiamen, the first three situated in the Guangdong Province) was announced in 1981. In 2010, this city hosted the 16th Asian Games. Historically, Guangzhou is famous for its openness to new possibilities. After 1978 it acts as a pioneer in China's reform, 'Guangdong is the province one step ahead in China and Guangzhou is the city one step ahead in Guangdong' (Xu and Yeh 2003, p. 361).

Guangzhou had a 12.9 million population in 2013, and covers an area of 7434.4 km^2 within 11 urban districts (Liwan District 荔湾区、Yuexiu District 越秀区、Haizhu District 海珠区、Tianhe District 天河区、Baiyun District 白云区、Huangpu District 黄埔区、Panyu District 番禺区、Huadu 花都区、Nansha District 南沙区、Zengcheng Distri ct 增城区、Conghua District 从化区). Liwan District, Yuexiu District, Haizhu District and Tianhe District are considered as urban areas while other districts have more characteristics of suburban areas.

According to the sixth National Census in 2010 statistics these four urban districts have a population of 5,046,575 which is 39.73% of the total population of Guangzhou, while the four districts only occupy 3.76% of the whole territory in this municipality (Zhang 2015). This research about urban redevelopment mainly concerns these four urban districts (Figs. 2.1 and 2.2).

2.2 China's Authoritarian Political Regime

Urban governance in China is deeply rooted in China's one-party political system. After the reform began in 1978, total control over the whole country has been relaxed; this change has been characterised as a transformation from a totalitarian regime to an authoritarian one (Nathan 2007). An authoritarian regime is described as a.

Political system with limited, not responsible, political pluralism, without an elaborate guiding ideology, but with distinctive mentalities, without extensive nor intensive political mobilisation, except at some points in their development. A leader, or occasionally a small group, exercises power within formally ill-defined limits but actually quite predictable ones (Linz 1964, p. 55).

These partially pluralistic elements are the most distinctive feature of an authoritarian regime compared with totalitarian regime and democratic one. This regime can involve some interest groups in the political process, which are decided by rulers rather than by institutional arrangement in a representative way. Ideology has less importance when practical considerations might be more dominant (Linz 2000).

After the Tiananmen Square protests in 1989, the authoritarian regime in China has fallen into crisis in terms of serious economic difficulties and legitimate problems.

2.3 Relation Between Guangzhou and the Central Government

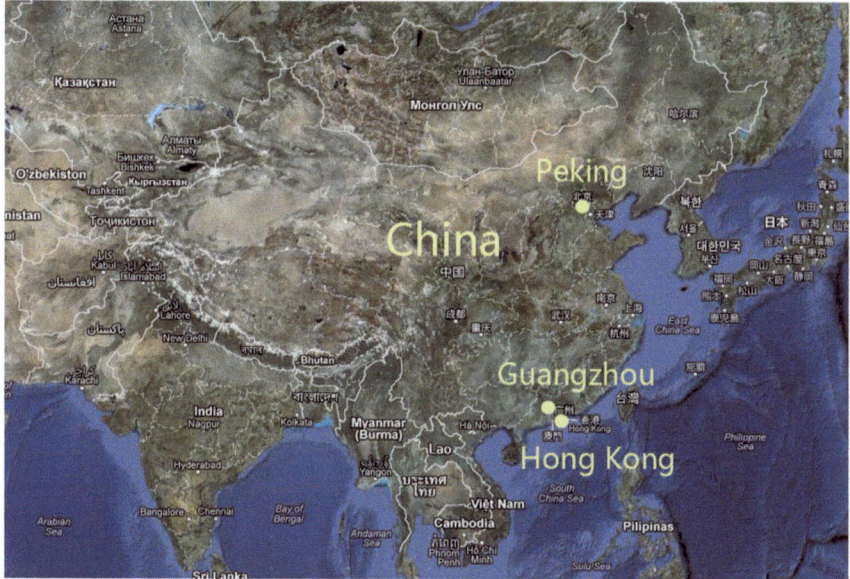

Fig. 2.1 The location of Guangzhou in East Asia. *Source* author's drawing based on Google map, 2016

However, the expected collapse of this communist regime has not happened after more than 25 years. Nathan (2003) raises the idea of authoritarian resilience to explain the survival of China's regime in those crises.

Authoritarian resilience in his context means four aspects of institutionalisation of the communist regime, namely a norm-bound succession process, the meritocratic promotion system among leaders, the differentiation between departments in government and more public participation in political processes. This concept has been developed by Cai (2008) in terms of the division of functions and duties between central and local authorities; this division aims to increase institutional resilience to face popular resistance and protests. The central state grants limited autonomy to the local state to deal with protests in local level; this autonomy means the local authorities could switch between concession and repression in different situations.

2.3 Relation Between Guangzhou and the Central Government

A flexible and effective central-local relationship is a crucial part in China authoritarian regime. The governmental relationship between Guangzhou and Beijing can be analysed in two aspects. The first is about distribution of duties, resources and authorities between two levels of government. This relationship has some features

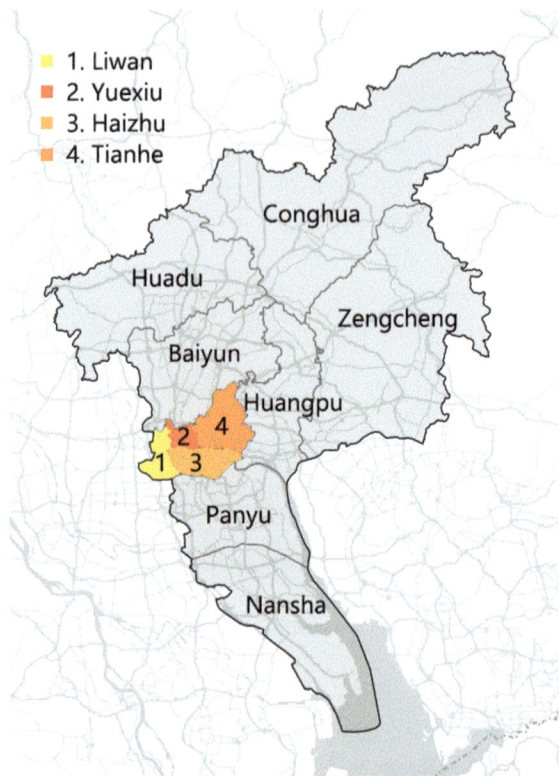

Fig. 2.2 Districts of Guangzhou in 2013. *Source* author's drawing

similar to the Federalist system. In such a system, the local leaders have strong incentives to develop the local economy and increase fiscal income because high-quality economic performance will increase their opportunities to be promoted to higher positions by the Beijing authorities. This is the second aspect of central-municipal relation. These two parts combine to form a pro-growth framework for Guangzhou.

2.3.1 Central-Local Relations and 1994 Reform

China, with its broad territory and diverse regions, has struggled with the principal-agent problem between central and local authorities. Since the 1980s, the central state established a contract system to define this relation. In such a system, the local government was promised a fixed amount of annual revenue to submit to the central state; besides this amount, every penny of public income belonged to the local government. These contracts stimulated a prosperous local economy and lead to the decline of central authority in the 1980s. This tendency threatened the national capacity to govern the whole country; therefore, the 'separating tax' reform was launched in

2.3 Relation Between Guangzhou and the Central Government

1994 (Wang 2013). The 1994 Tax Reform was based on market-preserving methods concerned with the distribution of power and assets, including two basic and related parts: a centralised fiscal system and a decentralised administrative system.

A centralised fiscal structure occupies the central position in the 'separating tax' reform. There are three sets of taxes distributed between Guangzhou and the central government. The first are central taxes, which include custom duties, personal and institutional income and consumption taxes, and taxes on profits from central state-controlled State-owned Enterprises (SoEs). The second category of taxes are central-local shared taxes, such as VAT (value-added tax) (25% goes to the local government and the rest (75%) to the central government) and securities' trading stamp duty. The third are local taxes, such as local enterprises taxes, individual income taxes, urban land use taxes and property and vehicle taxes. Basically, the central state grasps the most significant and convenient way to collect resources of taxes (Tao et al. 2009; He 2012).

To run such a reformed tax system, a newly established National Tax System with their local braches took charge of collecting taxes belonging to the central state. After the 1994 'separating tax' reform, the fiscal income of the central state increased rapidly; the authoritarian regime in Beijing has enforced its dominance by strengthening national financial abilities. However, the local government takes on the majority of the expenditure with less financial support. Table 2.1 displays the distribution between the Guangzhou government and central government. The central state possesses the majority of the fiscal income; however, the local expenditure is more dependent on the Guangzhou government. Therefore, a budget deficit becomes more and more serious in the municipal finance system.

Decentralisation of administrative resources is another part of the 1994 'separating tax' reform. This tendency was called 'localism' by Wu (1998); localism means that a decentralised decision-making process has transformed the local state from an executor of a central plan to a unit which could control the field of urban development, especially economic development. In fact, this Localism has a long history in the relationship between central and local states. Zheng (1995) described this relationship as a de facto federalism system when constitutionally China is definitely not a federalist country. Federalism here is a category of behaviour which helps to give provinces more autonomy and more opportunities to express their interests,

Table 2.1 Distribution of fiscal income between the central government and Guangzhou

Year	2017	2018	2019	2020
Total fiscal income of Guangzhou	5947.00	6205	6336.00	6155.8
Budgetary revenue of Guangzhou	1533.06	1632.30	1697.21	1721.6
Budgetary expenditure of Guangzhou	2185.99	2505.84	2865.12	2953.04
Fiscal transfer from Guangzhou to the central and provincial government	4413.94	4572.70	4638.79	4434.20

Source Guangzhou Survey Office of National Bureau of Statistics (2018, 2019, 2020); Wen (2021)
Note unit is one hundred million CNY

therefore, conflicts between different levels can be released. This is because activities of state are less controlled by law and constitution in China; therefore, some informal arrangements are necessary.

The central state has formulated the principles of policy, and the local ones produce detailed rules based on such principles. These activities are flexible and sometimes do not really obey rules of laws. Therefore, in economic aspects, the municipal level has significant autonomy in terms of decision-making in foreign investment, infrastructure, urban spatial structure and local enterprises. These autonomies help local authorities to solve the problem of budget deficits.

2.3.2 Leaders of Guangzhou in a Promotion Tournament

The central-local relationship after 1994 created the mismatch between financial resource and local expenditure; however, the increased autonomy (in theory) allowed local government the flexibility to respond to budget deficits and attempt to resolve them. This is the institutional level of problems. In an authoritarian regime, individuals at high levels are also influential because they exercise the most important authority in such a governmental system. To mobilise these leaders in local states to realise both economic growth and political stability, a Promotion Tournament has been launched in a non-election system. This model is based on the importance of incentives in transition and development.

The Chinese Communist Party state gears local leaders towards devoting attention and resources to economic development. This strong incentive is the connection between local economic performance and career turnover; this might explain why China's significant economic growth happens. This incentive and related turnover process has been described as a tournament model by (Li and Zhou 2005). A tournament model is a form of competition in which relative performance and ranking, not the absolute performance, will decide the result and the reward of competition. This model might encourage individuals to maximise their interests by realising designed achievements under neutral risks.

The promotion process in China's political system can be described as a tournament model due to several reasons: China's political system is a closed labour market without any exits for officials and competitive promotion is the only pathway to develop a career; the power for promotion is centralised at higher levels; the index to make the decision about turnover, such as the increasing ratio of GDP (Gross Domestic Product) or fiscal income, is objective and can be evaluated by the higher authority to decide who will be promoted in an information-asymmetrical environment; these indices can be compared between leaders in different regions; a joint venture between leaders is impossible. This is a typical situation for a tournament model.

Competition between different local states in terms of increasing the rate of GDP or fiscal income can benefit economic development in terms of protection with property rights and growth-bias policy for capitals. Innovation and experiment are encouraged

2.3 Relation Between Guangzhou and the Central Government

in this model, even if some of them are unofficial (at times, unlawful), for the reason that economic growth is a really crucial legitimacy problem for the Chinese government (Gries and Rosen 2004). Such experiments are realised by the autonomy of the local government to pursue innovation.

The problem with this tournament model is that the quest for economic growth ignores other aspects of governance, such as environmental problems and social conflicts. The debt crisis is another aspect of negative consequences associated with this model because the local state has strong incentives to employ all possible fiscal and financial tools and resources to achieve economic growth with less consideration for its results (Li and Zhou 2005; Zhou 2007). Jin et al. (2005) bring out another negative aspect of this model, the 'soft constraint' problem for local government. 'Soft constraint' means the local state has strong incentives to increase their expenditure to benefit local development. They may realise such expenditures through loans from the national bank system. It is difficult to control the desire of local leaders to borrow money to support urban development, even though such loans may surpass the ability of local governments to make repayments.

Xu and Wang (2010) modelled growth behaviour alongside two dimensions: competition effects and polarisation effects. Competition effects refer to the way in which successfully promoted local leaders spend more fiscal resources in the field of production than on consumption. The polarisation effect explains that leaders in relatively developed regions prefer to spend more public funding in support of industries, because improved economic conditions will lead to positive results in their career; such positive results might benefit their personal welfare in the future. However, leaders in relatively undeveloped areas might pay more attention to their own welfare directly. The diversity of China's regions results in local leaders using diverse strategies to maximise their personal interests.

The Promotion Tournament Model aims to explain China's significant economic development from a political economic perspective, through the changed focus from a political performance to an economic one. This model has been criticised by Tao et al. (2010) from the vision of a centralised power system. The higher leader will not employ such a clear index, such as the increasing rate of GDP, to make a judgement for promotion because this transparent method cannot match the essential element of the centralised system, the secret political process. Factions and personal relationships (patron-client relationships) might be more important in the process of promotion (Shih 2009). Even in a formal contract promotion is based on a comprehensive evaluation standard which includes moral requirements, political ability, diligence, performance and honesty. Economic performance is just one of the qualities required for promotion.

At the same time, official statistics are questioned in terms of their reliability, because the local officials have a strong tradition to 'create' numbers rather than purely record numbers (Tao et al. 2010). Similarly, Chen et al. (2016) also summarised that the tournament model has overlooked political aspects of promotion, such as political loyalty and factors of factions.

In such criticisms, Tao et al. (2010) argue that the ideas of Li and Zhou (2005) about the Promotion Tournament Model are based more on theoretical and statistical analysis; however, the promotion process in reality is complex and diverse in different situations, and more factors are at work in such a process. Leaders at the local level might face three dimensions of requirements for promotion: stability and legitimacy as the local state, economic performance as the local government, and maximising leaders' preferences in the hierarchical system (He et al. 2014). Such three dimensions may mix and conflict with one another in different fields.

However, even in these three parts, economic growth is still in the central position because stability and legitimacy are dependent on growth, and economic improvement has priority among the preferences of leaders. Moreover, the communist regime is glad to announce that economic performance is important for political promotion, for the reason that this announcement could increase the population's confidence about the regime's capacity to govern China (Chen et al. 2016). At least on the public dimension of promotion, economic performance in the leaders' tenure is an important factor. Therefore, the political leaders of Guangzhou, one of the largest cities in China, are also involved in this Promotion Tournament in terms of a pro-growth governance competition.

2.4 A Land-Oriented Growth Structure and the Role of Urban Redevelopment

Urban economic activities are dependent on urban land; the achievement of the pro-growth incentive for Guangzhou is also reliant on land issues. The main part of the growth mechanism is land-oriented due to the institutional structure around property rights and regulations relating to land, especially the local state's monopoly in the primary land market. The fees from leasing state-owned land to private users and taxes from the construction and transaction process provide strong support to the financial capacity of the local state, because since the 1994 tax reform, the income from land leasing and construction mainly belongs to the local government.

Urban redevelopment is an important part of this land-oriented growth regime, as property owners earn income from land leasing and construction, which arise during the redevelopment process. For instance, the 1998 amendment of the Land Administration Act announced that income earned from land designated as suitable for construction ought to be shared between the central and local government (30% belong to the central state, while 70% are local revenues). Existing construction land, already subject to urban redevelopment, could also produce fees and taxes which are totally controlled by the local state (He et al. 2014).

The land administration institutions mainly cover the administration of land property rights, the land administrative system and urban planning system; these create opportunities and set out the constraints and preferences of involved social groups in their social interactions in urban redevelopment. These three institutional elements combining with other institutional arrangements, such as housing reform, have built up an institutional structure to support land-oriented growth.

2.4.1 Land Property Rights and Regulations

Land resources are categorised by the state in terms of location, usage and ownership. In terms of different locations, land can be divided into urban land (cities, towns, industrial spots) and rural land. Before the Urban and Rural Planning Act in 2007, planning regulations were only applicable for building development on urban land. Land under diverse usages may include construction land (non-agricultural usage), agricultural land and unused land. Buildings may only be constructed on construction land; because of various types of ownership, land can be defined as state-owned land in urban areas and land in collective ownership in rural ones (Lin and Ho 2005).

These definitions of land have strong influences over processes concerning entitlement, usage and transactions of land. Such issues are controlled by the land administrative departments through territorial planning and other instruments. In addition, urban planning sectors are also crucial to realise the potential values of land when they regulate construction activities, such as function, density, height and architectural style of buildings. The functions of the land administrative departments and urban planning sectors are based on the property rights of land. Regulations from these two systems also influence how land property rights work in practice.

China's land ownership is special in terms of its communist characteristics and capitalist features. Its communist ideology used to control ownership and land use policy for decades. Such a control has been challenged significantly after reform since 1978, while the CCP preferred to focus more on economic development under the strategy of separating economics from politics (政经分离). Land started to be regarded as a crucial factor in production and it was felt that the distribution of lands should be more effective (Zhang 1997).

China's reform after 1978 is characterised by the phrase 'groping for stones to cross the river'; there is no well-designed 'blueprint' as an ideal destination of its transformation. A large number of experiments took place to explore the better pathway to develop as the 'stones' to cross the river, but there was a lack of clarity even about what would be on the other side of the river. The institutional changes to land ownership are typical cases of such methods: as there was no clear target of reform, vested interests stood in conflict with new requirements from the changed environment; this has become a common characteristic of China's reform.

At the beginning of the transition, foreign investors (including investors from Hong Kong and Taiwan) invested to build factories and properties on state-owned land or land in collective ownership. They felt unsafe because they had no property rights to the land which they had developed, as it belonged to the state in the case of urban land, or was collectively owned land in rural areas. They also felt uncertain about land use plans and land laws, unsure whether buildings on the land needed to conform to planning regulations.

To satisfy both the ideological requirements from Marxism and market conditions from foreign investors, political compromises were made in terms of property rights and land allocation methods. It meant that the use-rights of state-owned land were separated from state ownership; such land use-rights could be transferred to land users by payment of fees. The state is still the owner of urban land as the representative of the whole population of China.

At the same time, the former methods used to distribute land were still in force in which the state allocated land at no cost to work units or other publicly owned entities. These two opposing methods have typically accommodated the requirements of private investors, while (formally) maintaining the socialist principles. After several experiments involving land use-rights, transactions were established in Shenzhen and other Chinese cities, such a method has been approved by the first and second revising of the 1982 Land Administrative Act in 1988 and 1998 (the third revision took place in 2004) (Zhang 1997; Ho and Lin 2003; Lin and Ho 2005).

These differing methods for the conveyance of land produce a market track (in which land use-rights are bought) and a traditional, socialist land use track. The land management system is divided into a double track system. In the traditional socialist system, the free allocation process still exists for the use of land by state-owned departments or the non-profit sector. This type of land use rights is allocated by the state with no time limitation of expiry; but it cannot be directly transferred to other entities though the market process. It needs an expropriation process by the state or the payment of stipulated land fees to the local government.

The second land management system covers the conveyance of land for commercial functions by the private sector. This transaction is a market process conducted through negotiation or auction. Different categories of land usage have different lengths of use-right: commercial houses have 70 years, industrial land has 30 years and commercial functions, such as retail and hotel, have 50 years to use the lands. These lands can be sold to other users, or rented to others in the secondary land market, and such use-right land could be collateral to obtain bank loans. However, the primary land market, whether composed of freely distributed land or high-cost land conveyed by the purchase of land use-rights, is monopolised by the local state and operated by the land administration authorities (Zhu 2002; Lin and Ho 2005). In Guangzhou, these issues are controlled by the Guangzhou Land Development Centre which was established in 1992 as a department in the Guangzhou Municipal Land Resource and Housing Administrative Bureau.

State-owned lands in urban areas are controlled by the local government as the representative of the state, and the local government has the right to change the function of the land, consequently affecting local residents. The houses on the land belong to the householder, but they do not own the land under their house, they merely have the right to use it. Local government has the legal right and process to requisition state-owned lands; if there are public interests (there is great debate about its definition) in this requisition, compensation could be relatively low compared with the market prices of requisitioned houses. If this is a requisition for commercial interests, legally, the state should pay the same level of market prices to the former owners, or provide new houses within the same area (Zhu 2004).

Lands in collective ownership are located in rural areas and belong to the collective economic organisations, these organisations are owned by villagers, but if some of them leave this village and the collective economic organisation, they will lose their ownership of lands. Their land ownership can only be traded between villagers belonging to the same villages. Construction land in villages includes residential areas, township and village enterprises (TVEs) and infrastructure land. Legally, foreign investors could not construct on rural construction land; only urban construction land is suitable. The transformation of agricultural land to construction land is strictly controlled by the land administrative department in terms of the territorial plan and other institutions (Lin and Ho 2005; OECD 2010; Cheng 2011).

Conversion from rural land to urban land often happens in urban marginal areas, and sometimes in the city centre when urban villages are redeveloped; further consideration will be given to this issue later. This conversion is excellent business for the local state, because the activity is monopolised by local government and can produce significant revenue. According to the planning regulations imposed by land administrative institutions, construction land on rural land may be used for limited functions such as for public facilities and infrastructure.

The expansion of construction land plots depends on the increasing population of villages, requiring more (spacious) housing. Therefore, it is under strict control. However, the state-owned land in urban areas has much less restrictions on its development; therefore it has much higher value in the market. The local government can reassign rural land for development with compensation to the villagers, the former owners of rural collectively owned land. Compensation for such ownership changes sometimes are relatively low because they follow the standard of agricultural output from these lands; when the land is transformed into state-owned land with a greater range of uses permitted, the land value will increase significantly (Lin and Ho 2005; Cheng 2011; Wang 2013).

Such land conversion or change of use is profitable business for the local government to produce revenue: there is a great gap between prices of rural land and urban construction land, and the majority of income derived from the increase in price belongs to the local state. However, such a land conversion might threaten food security and social stability because the amount of agriculture land has been reduced and villagers have lost their means of production. Therefore, the central state has strict regulations over this conversion through the land administrative system (Lin and Ho 2005).

These regulations about land transactions between entities, transformation of agricultural land to construction land and conversion from rural land to urban land are executed by land administrative departments. They are about the land itself. In addition, urban planning is another regulatory mechanism to control construction activities on land. Urban planning activities are based on the 1989 Urban Planning Act, which has been developed into the 2007 Urban and Rural Planning (Xu and Ng 1998) argued that the ideology of socialist cities as a base for production has been changed; cities also act as important centres of transport, consumption and financial sectors after reform.

The function of planning is transformed from a project-lead plan to overseeing comprehensive development. Wu (1998) also stated that the philosophy of planning has been changed from political propaganda, to the stimulation of urban economic growth. However, in the central level of government, the Ministry of Construction has fewer concerns about growth, but more about the control of urban development in line with professional standards.

The statutory planning system contains a master plan, district plan and detailed development control plan (DDCP) according to the 1989 act (Xu and Ng 1998). DDCP is similar to zoning in the US system with a similar scale, function and objects to control the function, density, height and style of buildings; however, the DDCP is more about governmental commands, while zoning is about legislation. The majority of DDCPs are approved by the Guangzhou Urban Planning Bureau; some of the most important ones need to be submitted to the Guangzhou Municipal Government (Wu 1998). Urban redevelopment projects are mostly covered by DDCP rather than other categories of planning. DDCPs are often associated with governmental strategy to develop some specific areas and lease newly developed land or redeveloped areas in such locations.

Urban design is another mechanism for the local state to advertise some hot spot in the urban developmental map because urban design is non-statutory planning. This category of planning has less legal limitations and could conveniently follow the leaders' ideas about urban development. At the same time, the urban planning system also has the function of historical preservation which is usually an anti-growth institution in the urban redevelopment process. This preservation system has three levels: relic level (building level), district level and city level. The relic level is under the legal supervision of the Cultural Relics Protection Law of the People's Republic of China in 1982 and its amendments; urban planning. The other two levels are more concerned with specific historical preservation planning which is applied to involved areas with the normal DDCP. It is an additional regulation for historical districts or municipalities (Zhang 2010; Li 2012).

Guangzhou has been announced as a historical city with cultural significance in 1982; following which it was supposed to invoke planning for historical preservation. However, until 2014, Guangzhou has not yet applied protective planning measures for historical preservation because this might be an obstacle for economic development in old towns (Interview from Li 2013).

Based on such urban planning systems in Chinese cities, there are two innovations in Guangzhou. The first one is the Planning Commission which has been built up since 2006. This commission with members composed of city leaders, leaders from involved bureaus, experts about planning and representatives from citizens, has the administrative power to approve any important planning issues. All major modifications of DDCPs in redevelopment projects need to be permitted by this commission. The mayor of the commission becomes more and more dominant over other members, in spite of its diverse and varied composition (Interview from Li 2013, 2014; Interview from Yuan 2014).

Another important institutional change is the strategic development plan at the city level to adopt the economic competition between Chinese cities. The new type of plan is a non-statutory plan to deal with limitations of the master plan, which has a lengthy process of approval and does not focus on attracting investment to benefit industries. The Guangzhou strategic development plan is the first one among Chinese cities and aims to develop a strategy to adopt its extended administrative territory in 2000 by grasping new opportunities.

This plan is more concerned with the intention of the local leader (in the case of Guangzhou this is the mayor) to develop the urban economy. It might lead to less consideration about social and environmental issues with scientific approaches to planning. This plan employs a spatial strategy which includes the location of new industrial clusters, new public transport systems and a new polycentric urban structure (Wu and Zhang 2007). Urban redevelopment projects are involved in such strategic development as part of spatial strategy and mega projects.

2.5 An Institutional Framework of Land-Oriented Growth and Redevelopment

The institutions discussed in this chapter bring about fundamental interaction processes between central and local government, which means power, resources and responsibilities are distributed and redistributed between different administrative levels. Within such an authoritarian regime with top-down political control, partial autonomy of local state and limited social participation, a great number of motivations for local development and competition are produced in an inner-city competition.

At the same time, the central government gets the major part of the tax income and funds a minor part of the social services, so the local government needs to find new financial resources to support this unbalanced structure. Furthermore, according to the 'Promotion Tournament Model', the competition for political promotion is based on competition of economic performance in the Chinese bureaucratic system. Increased GDP and revenue are the main indicators for economic performance (Li and Zhou 2005).

Meanwhile, local fiscal income can stimulate economic growth through constructing infrastructures and public facilities to support and upgrade local industries. Therefore, the competition of development between local governments and officers make the consequence of unbalanced distribution more serious. During the same time, reform of the housing system stopped the public supply of housing; people need to find housing in the real estate market. The real estate market is built on the basis of the land market, in which the government monopolises the primary local land market, which means local governments could get monopolistic rents to support the system of tax distribution and develop the urban economy to win the competition of development.

Therefore, there is a land-oriented growth pattern in Chinese cities, which depends on consuming land resources to stimulate economy and produce fiscal income. Compared with supporting economic growth, land market issues directly contribute to increase local financial resources. Property rights of land and regulations on land provide opportunities for the local state to grasp revenues.

Commercial constructions may only legally be built on urban construction land. There are two pathways to gather urban construction land: through expropriated urban land or through conversion from rural land. Both of these processes can only be operated by land administrative departments because the transformation of ownership, usage and category is controlled by these departments. Governmental sectors' monopoly of land as the only buyer brings about strong abilities to demand a low price. After the acquisition of land from former users, the process of urban planning is then used to increase land value.

Urban planning as a professional tool is mainly controlled by the local government to maximise the values of land to be leased in the land market. DDCP, urban design or strategic plans, all of them can contribute to increase land values in terms of providing a clear developmental strategy, reducing uncertainty in the future and advertising the project and the city.

As for the results, firstly, the local state can benefit from the land-let market when it is the only seller in the primary land market, whatever the land is, from expropriated urban land or conversion from rural land, and whoever they are sold to, private entities or public enterprises. This land-let income is more dependent on the real estate industry which has developed significantly since 1994.

Secondly, the local state can get tax income from the construction process and selling process of the real estate industry because most of these tax bases belong to the local state, according to the reform of the system of tax distribution in 1994.

Thirdly, the local government can get loans from the national bank system by using the state-owned lands under the effective control of the local government and their future income as mortgages. Such income and loans will be invested in infrastructure to develop urban economy (Tao et al. 2009). Urban redevelopment has an important role in such a process because it can gain fiscal income through the three methods mentioned above; moreover, the land income from land redevelopment mainly belongs to the local state.

Such institutions discussed above have interconnected with each other on the issues of developing or redeveloping land to produce taxes and fiscal incomes. The 'separating tax' reform after 1994 and the Promotion Tournament bring about incentives to local leaders to stimulate the growth of economy and revenue; housing reform has released the potential of the real estate market to increase the attraction of the urban land market; highly state-controlled property rights of urban and rural land give the local state dominance to realise the potential values of land; land administrative sectors have institutional resources to operate the land-value increasing process before the primary land market; urban planning is a professional tool to organise land-leasing activities and maximise land values.

Of course, this is not the whole story of urban economic growth in Chinese cities; developments in industrial zones have some different logic (Tao et al. 2009). However, it is an institutional framework to realise incentives of local leaders to stimulate economic growth with successful results in the last few decades. Urban redevelopment is a crucial part of such growth because redevelopment has been involved in every institutional activity focusing on growth: it produces land rent; it increases taxes from construction and selling; it supports the upgrading of the urban economy; it facilitates mortgages to provide loans for building infrastructures. It is a pillar of this land-oriented growth.

References

Cai YS (2008) Power structure and regime resilience: contentious politics in China. Br J Polit Sci 03(38):411–432

Chen J, Luo DL, She GM, Ying QW (2016) Incentive or selection? A new investigation of local leaders' Political Turnover in China. Soc Sci Quart 1–19

Cheng XY 程雪阳 (2011) Zhongguo tudizhidu defansiyubiange 中国土地制度的反思与变革 [The reform of land system in modern China: review on the perspective of public law]. Ph.D Thesis. Zhengzhou University 郑州大学

Gries PH, Rosen S (2004) State and society in 21st century China: crisis, contention and legitimation. Routledge Curzon, London

Guangzhou Survey Office of National Bureau of Statistics (2018) 2017nian guangzhoushi guominjingji he shehuifazhan tongjigongbao 2017 年广州市国民经济和社会发展统计公报 [Guangzhou statistical yearbook of economic and social development]. [online] Available at: https://gdzd.stats.gov.cn/gzdcd/gz_tzgg/201804/t20180409_169467.html. Accessed: 25 May 2023

Guangzhou Survey Office of National Bureau of Statistics (2019) 2018nian guangzhoushi guominjingji he shehuifazhan tongjigongbao 2018 年广州市国民经济和社会发展统计公报 [Guangzhou statistical yearbook of economic and social development]. [online] Available at: https://gdzd.stats.gov.cn/gzdcd/gz_tzgg/201904/t20190403_169468.html. Accessed: 25 May 2023

Guangzhou Survey Office of National Bureau of Statistics (2020) 2019nian guangzhoushi guominjingji he shehuifazhan tongjigongbao 2019 年广州市国民经济和社会发展统计公报 [Guangzhou statistical yearbook of economic and social development]. [online] Available at: https://gdzd.stats.gov.cn/gzdcd/gz_tzgg/202005/t20200526_176221.html. Accessed: 25 May 2023

He SJ (2012) two waves of gentrification and emerging rights issues in Guangzhou, China. Environ Plan A 44(12):2817–2833

He YL 何艳玲, Wang GL 汪广龙, Chen SG 陈时国 (2014) Zhongguo chengshizhengfu zhichu zhengzhifenxi 中国城市政府支出政治分析 [A political analysis of the expenditure of municipal governments in China]. Soc Sci China 中国社会科学 7:87–106

Ho SPS, Lin GCS (2003) Emerging land markets in rural and urban China: policies and practices. China Quart 175:681–707

Jin H, Qian Y, Weingast BR (2005) Regional decentralization and fiscal incentives: Federalism, Chinese style. J Public Econ 89(9–10):1719–1742

Li C (2012) The end of the CCP's resilient authoritarianism? a tripartite assessment of shifting power in China. China Q 211(211):595–623

Li HB, Zhou LA (2005) Political turnover and economic performance: the incentive role of personnel control in China. J Public Econ 89(9–10):1743–1762

Lin GCS, Ho SPS (2005) The state, land system, and land development processes in contemporary China. Ann Assoc Am Geogr 95(2):411–436

Linz JJ (2000) Totalitarian and authoritarian regimes. Lynne Rienner Publishers, Colorado

Linz JJ (1964) An authoritarian regime: Spain. In: Allardt E, Littunen Y (eds) Cleavages, ideologies and party systems: contributions to political sociology. Transactions of the Westermack Society, Helsinki

Nathan AJ (2003) Authoritarian resilience. J Democr 14(01):6–17

Nathan AJ (2007) Congjiquantongzhidaorenxingquanwei: zhongguozhengzhibianqianzhilu 從極權統治到韌性威權—中國政治變遷之路 [Political change in China: from totalitarian rule to resilient authoritarianism]. Translated by He, D.M. 何大明. Liwen wenhua, Taipei

OECD (2010) OECD territorial reviews OECD territorial reviews: Guangdong, China 2010. OECD Publishing, Paris

Shih VC (2009) Factions and finance in China: Elite conflict and inflation. University Press, Cambridge

Statistics Bureau of Guangzhou Municipality and Guangzhou Survey Office of National Bureau of Statistics (2014) 2014 guangzhou tongjinianjian 2014 广州统计年鉴 [Guangzhou statistical yearbook, 2014]. China Statistic Press, Beijing

Tao R 陶然, Lu X 陆曦, Su FB 苏福兵, Wang H 汪晖 (2009) Diqujingzhenggeju yanbianxiade zhongguozhuangui: caizhengjili hefazhanmoshi fansi 地区竞争格局演变下的中国转轨: 财政激励和发展模式反思 [China's transition and development model under evolving regional competition patterns]. Econ Res J 经济研究 44(07):21–32

Tao R 陶然, Shu FB苏福兵, Lu X 陆曦, Zhu YM 朱昱铭 (2010) Jingjizengzhang nenggoudailaijingshengma? Duijingshengjinbiaojingsaililun deluojitiaozhan yushengjishizhengchonggu 经济增长能够带来晋升吗?——对晋升锦标竞赛理论的逻辑挑战与省级实证重估 [Can economic growth lead to promotion? Logic challenge and provincial empirical revaluation to promotion tournament theory]. Manage World 管理世界 12:13–26

Wang H 汪晖 (2013) Zhongguo zheng di zhi du gai ge : li lun, shi shi yu zheng ce zu he中国征地制度改革: 理论、事实与政策组合 [Land requisition system reform in China: theories, facts and policy portfolio]. Zhejiangdaxue chubanshe, Hangzhou

Wen GH (2021) 2021nian guangzhoushi renmingzhengfu gongzuobaogao 广州市人民政府工作报告 (2021年) [Guangzhou municipal government annual report 2021]. [online] Available at: https://www.rd.gz.cn/zyfb/bg/content/post_213271.html. Accessed: 25 May 2023

Wu FL (1998) Urban planning system in China. Prog Plan 51:169–254

Wu FL, Zhang J (2007) Planning the competitive city-region: the emergence of strategic development plan in China. Urban Aff Rev 42(5):714–740

Xu XX 徐现祥, Wang XB 王贤彬 (2010) Renmingzhi xiade guanyuan jingjizengzhangxingwei 任命制下的官员经济增长行为 [Growth behavior in the appointment economy]. China Econ Quart 经济学季刊 9(04):1447–1466

Xu J, Ng MK (1998) Socialist urban planning in transition: the case of Guangzhou, China. Third World Plann Rev 20(1):35–51

Xu J, Yeh AGO (2003) City profile: Guangzhou. Cities 20(5):361–374

Zhang XQ (1997) Urban land reform in China. Land Use Policy 14(03):187–199

Zhang J 张杰 (2010) Congbeilun zouxiang chuangxin: chanquanzhidu shiyexiade jiuchenggengxinyanjiu 从悖论走向创新: 产权制度视野下的旧城更新研究 [From paradox to innovation: research about urban regeneration based on the perspective of property rights]. Zhongguo jianzhugongyechu banshe, Beijing

Zhang HY 章鸿雁 (2015) Chanye shengji daoxiangxia de jiuchengqu gengxin jianshe- yi guangzhoushi weili 产业升级导向下的旧城区更新建设——以广州市为例 [Urban regeneration oriented by industrial upgrade: a case study of Guangzhou]. Zhonghuajianshe 中华建设 2015(1):82–85

Zheng Y (1995) Institutional change, local developmentalism, and economic growth: the making of semi-federalism in reform China. Ph.D Thesis. Princeton University. [online] Available at: http://ezproxy.princeton.edu/login?url=http://search.proquest.com/docview/304222974?pq-origsite=summon. Accessed: 22 Oct 2016

Zhou LA 周黎安 (2007) *Zhongguo difangguanyuan de jingshengjinbiaosai moshiyanjiu* 中国地方官员的晋升锦标赛模式研究 [Governing China's local officials: an analysis of promotion tournament model]. Econ Res J 经济研究 42(07):36–50

Zhu J (2002) Urban development under ambiguous property rights: a case of China's transition economy. Int J Urban Reg Res 26(1):41–57

Zhu J (2004) From land use right to land development right: institutional change in China's urban development. Urban Stud 41(7):1249–1267

Open Access This chapter is licensed under the terms of the Creative Commons Attribution 4.0 International License (http://creativecommons.org/licenses/by/4.0/), which permits use, sharing, adaptation, distribution and reproduction in any medium or format, as long as you give appropriate credit to the original author(s) and the source, provide a link to the Creative Commons license and indicate if changes were made.

The images or other third party material in this chapter are included in the chapter's Creative Commons license, unless indicated otherwise in a credit line to the material. If material is not included in the chapter's Creative Commons license and your intended use is not permitted by statutory regulation or exceeds the permitted use, you will need to obtain permission directly from the copyright holder.

Chapter 3
Three Phases of Urban Redevelopment Governance

Abstract This chapter aims to describe what happened in the field of urban redevelopment in Guangzhou from 1990 to 2015. The patterns of governance in this period can be divided into 3 phases; different phases have quite distinct characteristics of social cooperative modes and various state-market-communities relations. It comprises 3 phases with different patterns of governance: first, the Primitive Market Phase (1990–1998); second, the Pure Government Phase (1998–2006); and third, the Multiple Players Phase (2006–2015). These phases are strongly connected to periods of control by the mayors of Guangzhou, especially in the first two phases. However, the modes of governance were not changed immediately after the changes of mayors; every new leader needs time to adapt a new position, control new resources, modify existing policy and build up new agendas. There are always delays between the announcement of new mayors and establishment of a new mode of governance. This research prefers to identify phases based on the period of leadership of mayors because this method is simple and clear, and emphasises the importance of leaders in governance. In every phase, modes of social interactions and the state-market-communities relations will be expressed.

Keywords Urban redevelopment · Governance · Mayor · State-market-communities relation

3.1 Primitive Market Phase (1990–1996)

This section of the chapter provides a narrative about urban redevelopment in Guangzhou from 1990 to 1996. This phase started in 1990 when Mr. Ziliu Li became the acting Mayor of Guangzhou (this position holds the same authority and duty as the mayor but awaits legal approval from the People's Congress of Guangzhou). This phase is named as the Primitive Market Phase because of two reasons. On the one hand, the governance mode in this phase was a semi-market mode. On the other hand, market entities were dominant in this phase in social interactions in redevelopment.

3.1.1 Overall Characteristics of Redevelopment in This Phase

Before 1990, there was a 'six-uniformed' system in the urban redevelopment process, in which the government administered standardised planning, compensation, design, construction, facility and management elements in redevelopment projects (Huang 2013). Under this 'six-uniformed' system there are two categories of redevelopment, the redeveloped declining residential areas for existing citizens, and the redeveloped commercial residential quarter selling to customers outside mainland China (Kong 2008). This system is a government-dominant mode of redevelopment, which has been abandoned by Mr. Li since 1990. He set up a new system which combined a land distribution process based on the personal decision of the leader, with the autonomy of developers in the whole redevelopment process.

> From the 1980s, Guangzhou aimed to build up a 'six-uniformed' system, which included a uniformed plan and development. However, Mayor Li totally abandoned such a system after the start of his tenure as mayor. Whatever you wanted to do, you needed to speak to Mayor Li. If he agreed with you, you got permission for your project without any other administrative permission (such as urban planning permission). In this situation, most cases of redevelopment are chaotic [out control of governmental planning], except the Jinhua Street project and Dongfeng Street project, which are still in the authorised channels.
>
> (Ye, senior officer, interview, 12/2013).

This phase is named the Primitive Market Phase because of its free market style governance mode and the influential roles of developers in redevelopment projects. This market was active in terms of the large scale of investment and huge numbers of projects. The formal redevelopment process had not been established. Governmental regulation was weak while developers held dominant roles in this phase. The infrastructure-driving redevelopment was the main approach for the local authority to propose redevelopment projects. Historical preservation was in danger in this phase as this held a low priority.

(1) **Redevelopment as a Hot Spot**

In this stage, urban redevelopment had entered maybe the most active period in the history of redevelopment in Guangzhou. Yuan (scholar and planner, interview, 2014) insists that there was more than 2,000 real estate companies active in the early 1990s; 2,208 redevelopment projects had been approved between 1992 and 1998. Parts of the projects had not started until 2006 (Pan et al. 2006). Between 1992 and 1996, 175 km^2 of land had been freely distributed to developers. Some developers did not possess the capacity to redevelop such plots of lands; they just wanted to sell the land. Therefore, a huge amount of land had not been used until 2012. During the same period, the Guangzhou government had announced 1,194 notices about removal and demolition.

(2) **Informal Process of Redevelopment**

Redevelopment did not have a defined formal administrative process in this phase. There was no need to gain specific permission from the urban planning and land administrative sectors as all the developers needed to do was obtain the agreement of the local leader. This negotiation has more characteristics as a formal one.

Some developers obtained land through their personal relationships with city leaders or politicians at higher levels of government. In some of these projects land fees for land leasing were not paid until the real estate properties were sold after a few years. Developers could occupy new plots when they get agreement from the mayor without permission from other administrative departments or legal processes. Following the land transaction their permission for construction included just a red line of plot, without any limitation of plot ratio and height of buildings. Sometimes the ownership of land was still unclear when it was sold or distributed to developers. Some of the projects have not been built after several years.

> He [Mr. Li] provided so many opportunities to developers who had strong connections with local authorities. These developers came to me with the personal permission of local or central leaders; they told us they were planning to develop Guangzhou. I asked them [some officers] whose permission these developers had got; they told me everyone [in high positions of China's Communist regime].
>
> (Yuan, scholar and planner, interview, 01/2014).

(3) **Waived Regulations**

In this stage, urban planning was just a block plan to control redevelopment; however, a block plan is a relatively weak regulation. At the same time, Guangzhou had been listed as a 'Famous Historical and Cultural City' in 1982; this also had less impact to regulate redevelopment activities because such recognition did not bring about related legal or administrative methods to protect the historical and cultural heritage in old towns. As a result the scale, density and height of redevelopment were limited by few regulations. The developers had more freedom to maximise their interests.

From Liu's (2006) perspective, local political leaders are eager to attract investment, especially from Foreign Direct Investment (FDI), into the field of urban redevelopment. To achieve this the whole planning system of regulations in the 1980s, such as laws, rules, standards and the requirements of other involved institutions, were abandoned (Ye 2014, p. 162). Because of waived regulations developers sought to increase the plot ratio in every case, but decrease social services at the same time (Wang 2010).

> At that stage, there was not a strict control in historically and culturally sensitive areas. Some projects could not happen under contemporary regulations. At that phase there was just a useless block plan, which was very basic.
>
> (Huang, urban planner, interview, 01/2014).

> China has built up a list of 'Famous Historical and Cultural City' [to protect historical and cultural areas] since 1998 [in fact since 1982]. However, it just displayed some principles; other things were unclear, such as what should be protected? In which degree they should be preserved? There were no lawful requirements, nor required in planning. Therefore, even though Guangzhou was among the first few cities to have entered the first list of 'Famous Historical and Cultural City' there was no consensus in terms of historical and cultural preservation.
>
> (Ye, senior officer, interview, 12/2013).

(4) Dominant Role of Developers

Urban redevelopment was mainly supported by private capital. In 1994, 32.22 billion CNY had been invested in redeveloping Guangzhou, of which 15.04% came from Chinese banks; 43.03% was contributed directly by developers; and 30.4% was produced by pre-payment of sold properties from buyers (Wei 1997). From 1992 to 1996, 418,740 households were removed and demolished, of which only 7.792% of the families displaced lost their homes because of public projects. The remaining number of properties, 92.208%, were demolished to make way for private projects (Lin 2012).

In the interaction between government, developers and communities, the developers have occupied dominant roles. They controlled the choice of projects, when and how to redevelop the site, found resources to support projects and shared the majority of profits in the projects. For instance, in the Ximenkou Square redevelopment project, the developer had a primary contract with the local state in terms of distribution of profits; it announced that the developer could obtain 65% and the government receive 35%. However, the developer was unsatisfied and renegotiated with the government to change the conditions of the contract, otherwise they would not invest. The result was that the government made a concession and agreed that the developer could possess 89% of the profits (Wang 2009).

(5) **Spatial Strategy: Infrastructure-Driving Redevelopment**

Based on the marketing-dominate strategy, redevelopment in Guangzhou employed an infrastructure-driving and small-scale spatial strategy; this started a new redevelopment project which was a new proposal for improving the conditions of a road or building underground. Because the local authority lacked funding during this stage, they had to sell some land beside this proposed road to potential developers. The lands designated for sale were divided into small plots before transactions. Developers expected increased land values in the sites sold due to the very possibly improved traffic conditions in the future. Huang (urban planner, interview, 2014) called it 'a perfect match between developers and government, between free enterprise mode and road driving spatial strategy'.

(6) **High Density of Redevelopment and Threats to Historical Preservation**

The Primitive Enterprise Phase brought prosperity in urban redevelopment; developers had more freedom, and thus more incentives to engage in redeveloping. To maximise their interests with the waived regulations of urban planning, a high density of redevelopment projects were popular in this phase. This is a threat to the traditional landscape in historical areas. It also brought about high pressure on public facilities such as water supply, waste treatment and traffic. Mayor Li has admitted that in this phase historical buildings, such as Daxiao Mazhan and traditional colleges had been destroyed (Zhang 2008). Mr. Shi, the former leader of the Urban Planning Bureau of Guangzhou, also thought historical conservation was problematic in the Line One Underground Project because developers were dominant in this redevelopment (Chen and Li 2007).

3.1.2 Typical Project: Liwan Square Redevelopment (1992–1995)

The Liwan Square Redevelopment Project was the most significant redevelopment project in the 1990s (Jiao 2010; Liu and Fang 2010). This project has been mentioned many times in Guangzhou from a negative perspective (Yuan and Xie 2010). Although there is less information about this case because it happened 20 years ago, it still has a strong impact both on the real landscape in the core of the old town, and in people's minds who are interested in urban redevelopment.

> After that there was an expansion of numbers of redevelopment in old towns, which included the typical case, Liwan Square.
>
> (Ye, senior officer, interview, 12/2013).
>
> Therefore, the Liwan Square project provided an impetus for the origin of strategic planning in Guangzhou at 2000.
>
> (Y2, scholar and planner, interview, 01/2014).

The Liwan Square Redevelopment Project was located close to Shangxiajiu Pedestrian Street, a popular traditional shopping space in Guangzhou. This project has used 4.5 ha of land to build 140,000 m^2 of market and more than 1,000 flats. This project was organised by Mr. Hua Huang, and his South International Group, an enterprise registered in Hong Kong. Liwan Square was located in a rundown traditional commercial area with a complex ownership situation.

In 1992, Mr. Huang attended the introduction of the Liwan Project organised by the Guangzhou government in Hong Kong; the tension between the excellent location and difficult ownership conditions was high (Zhang et al. 2010; Honour 2010). After negotiation, the South International Group and Guangzhou Land Resource and House Management Bureau (广州市国土房地产管理局) cooperated to establish the Suihua Real Estate Enterprise to redevelop this project. In this company 100 per cent of the investment came from the South International Group while the Bureau contributed the land for this project as an investment under the name of Ruihua.

After the project was completed, 165,000 m^2 of flats and 30,000 m^2 of shops belonged to the South International Group; 15,000 m^2 of shops and 30,000 m^2 of parking space were distributed to Ruihua (to the Bureau actually) (Liu 2008b).

An underground stop close to Liwan Square was also an important driving force for the redevelopment. According to the developer's statement they took less than two years to demolish more than 2,000 households, 30,000 shops and 5 factories, and remove more than 8,000 relevant to this redevelopment. The developers have provided incentives for both the owners of demolished properties and the company undertaking the demolition work to cooperate with the developers. The demolition cost around 800 million Hong Kong dollars and 1.2 million was spent in construction and marketing. In 1996 the project was completed (Law of China Net 1993).

After handing the new built properties over to new owners a large number of legal arguments took place. Hundreds of buyers from Hong Kong were unsatisfied with this project in terms of two main issues. Firstly, the buyers received their properties three months later than the date in their contract with developers; secondly, the developers had changed the design of properties which had reduced the value of the properties (Liu 2008a).

Between 1996 and 2001 the buyers took legal action about the two issues, prosecuting the South International Group more than 500 times in different courts (Law of China Net 2001). For instance, on 5 February 1999 in the High Court of Guangdong Province, the prosecution counsel claimed that the developers have cheated buyers in their advertisement of the project, providing properties with a lower specification than previously advertised. However, the lawyers of the developers answered that the advertisement was in Hong Kong newspapers which were outside the duty of the High Court of Guangdong Province. In addition, the developers' lawyers argued they have official permission to delay this project for three months.

In response, the prosecution lawyers insisted that permission was only reached between the developers and government, not between developers and buyers. After arguments, both sides agreed to make a compromise with each other (Law of China Net 1999). Such lawsuits have continued until Mr Hua Huang was bankrupted in 2001 (Wu 2009). The buyers involved in the disputes suspected that the Guangzhou Land Resource and House Management Bureau had supported Mr. Hua Huang; therefore, they could not win the prosecutions (Liu 2008a, b) (Figs. 3.2 and 3.3).

Economically, the Liwan Square redevelopment might be a successful project. However, it is famous for its negative influence on historical preservation in the city

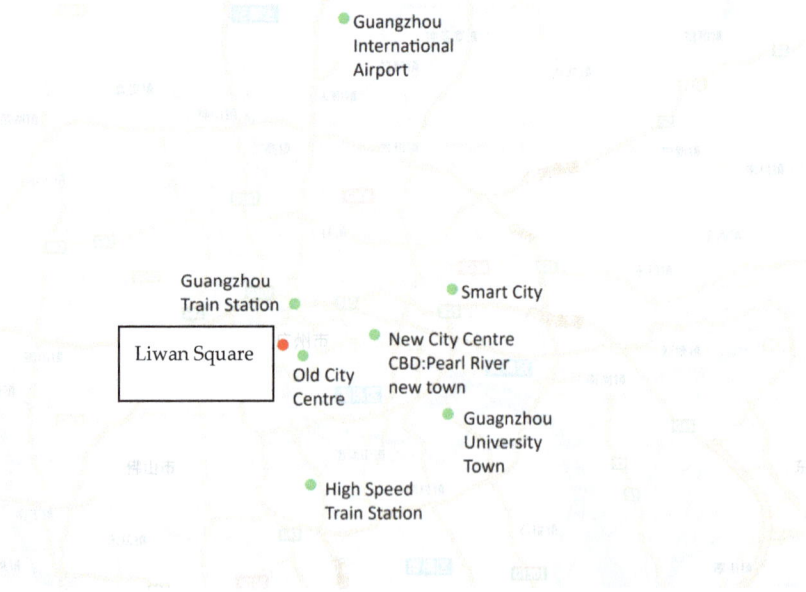

Fig. 3.1 Location of Liwan square redevelopment project. *Source* author's drawing based on http://map.baidu.com/, 2016

3.1 Primitive Market Phase (1990–1996)

Fig. 3.2 Master plan of Liwan Square (A1 is before redevelopment; A2 is after redevelopment). *Source* Chen 2014

Fig. 3.3 View of Liwan Square. *Source* author 2015

centre. Many scholars and planners criticised its consequences, such as removing the original communities, destroying the traditional urban context and interrupting the local landscape (Zhu 2003; Peng 2007). This failure in historical conservation was related to the weak urban planning regulations which were in operation during this phase.

> Mayor Lin [the mayor following Mayor Li] recognised that in urban redevelopment, projects such as Liwan Square cannot happen again….So ….
> (Yuan, scholar and planner, interview, 01/2014).
> If you allow capital investors to do whatever they want, Liwan Square will appear everywhere.
> (Wang, scholar and planner, interview, 12/2013).

3.1.3 Summary of the Primitive Market Phase

This phase is described as the Primitive Market Mode, which was close to the market mode of governance in terms of several ways:

There were thousands of small developers with thousands of small projects supported by private capital in the field of redevelopment; it was a typical landscape of a primitive capitalist market with less monopoly; redevelopment activities were less regulated; enterprises had more autonomy to make their own decision; interactions between government and developers were more dependent on relative advantages rather than hierarchical orders; the contract is the main method to distribute interests between entities from the state, market and communities; negotiation is the main mechanism to solve problems, in some cases, legal methods were also employed. These characteristics could match features of the market mode as an ideal mode of governance. In some special cases, the government had invested in these projects; the market had less dominance but still occupied an important position.

3.2 Pure Government Phase (1998–2006)

This is the second phase of urban redevelopment analysed in this study. It was famous for the clear announcement from Mayor Shusen Lin: all private developers were forbidden to enter the field of urban redevelopment in Guangzhou after 1998 (Huang 2013). This obvious anti-market policy was applied for several years to totally change the landscape of redevelopment.

3.2.1 Overall Characteristics of Redevelopment in This Phase

In this phase, urban redevelopment projects were established according to political agendas, such as political performance and legitimacy requirements. Public funding was the main resource to support redevelopment activities. Profits were not the priority in individual projects; public interests were considered much more than in the previous phase. During these years land management was reorganised as Mayor Lin had recognised the disadvantages of the Primitive Market Phase, such as disputes over compensation, delayed projects and redevelopment that threatened the landscape of the city centre (Deng 2005).

The governance style in this phase was mainly displayed by Mayor Lin's attitude to developers. He never attended any dinner organised by developers and never appeared in any developers' projects. He always thought that developers aimed to maximise their economic interests rather than develop an excellent metropolis. Mayor Lin did not approve any negotiated land-leasing project which was a cheap approach to win land and had been operated in a large number of projects in Mayor Li's period (Xiang 2006). Mayor Lin is reported to have made jokes with developers in the Fourth session of the Twelfth National People's Congress in Guangzhou in 2006. He mentioned that it was lucky for him that developers had earned a huge amount of money in the last ten years, otherwise, developers would hate him too much because he had forbidden them from obtaining land from negotiation and from redeveloping old towns in the city centre (Yu et al. 2006).

3.2.2 Typical Project: Three Changes of Guangzhou

The 'Three Changes' Redevelopment aimed to comprehensively transform the image of the whole of Guangzhou in three stages. The first proposed redevelopment was a small change over one year to demolish illegal buildings. The second aimed at medium-term changes over three years to improve traffic conditions (traffic infrastructure), public squares and appearances of buildings along main roads. The third and most ambitious proposals aimed at long-term changes until 2010 by which Guangzhou should have been transformed into a modern metropolis (Lin 2013).

On 31 July 1998, the leaders of Guangdong Province had organised a conference about urban construction in Guangzhou. In this meeting, Mr. Changchun Li, secretary of the Guangdong Provincial Communist Party Committee as the highest political leader in Guangdong Province, had proposed the 'Three Changes' Redevelopment of Guangzhou. On 29 and 30 October 1998 in response to the proposals from the higher political level, Mayor Shusen Lin had signed agreements between the municipal government and district government and other departments; within these agreements the targets and responsibilities in the small-change redevelopments had been defined (Wang et al. 2000).

The three-year medium-term changes (1998–2001) were the focus of the 'Three Changes' Redevelopment because 2001 was the year to host the 9th National Games, and it was important for Guangzhou to present achievements of redevelopment. In addition, 2010 was outside Mayor Lin's term of office; the policy about the 'Three Changes' Redevelopment might be changed.

Efforts to improve the urban image under the three-year medium-term changes focused on 109 projects. These 109 projects had originated in 1997 when Mayor Lin had required the General Director of Guangzhou Urban Planning Bureau, Mr. Feng Dai, to design projects to improve the urban image. It was decided that this project should spend relatively less funding to impress citizens in terms of changing the image of Guangzhou. In March 1998, 109 individual projects that aimed to upgrade urban appearances had been identified by the Urban Planning Bureau to Mayor Lin (2013).

The redecoration of Beijing Pedestrian Road, in one of the first group of 109 individual projects, was constructed between May and September 1999 under the supervision of the Yuexiu District Government. This project took only 123 days to finish because it aimed to celebrate the 50th anniversary of the People's Republic of China on 1 October 1999.

To accelerate such a fast-track construction, the Management Office of Beijing Pedestrian Road, led by district leaders, cooperated with various departments and construction companies. Shangxiajiu Pedestrian Road was another important project in the image improvement projects. These two projects cost 66 million CNY of public funding. Before the 9th National Games hosted in 2001, 2.16 billion CNY governmental money had been spent in redecorating more than 6,000 buildings among main roads (Liu 2016).

Garden Square was one special case in improvement projects in the three-year medium-term changes. Proposals for the square had been made in 1998 to create a public space in Huanshidong Road with a budget of 300 million CNY from the Municipal Construction Committee. In 1999, a design competition for the square had produced several plans to build it. One of the proposals planned to construct the square between July 1999 and August 2000 with a budget of 388 million.

This design was attractive to the Guangzhou government; however, Mayor Lin and other leaders no longer supported the project. This was due to several reasons. Firstly, the budget set aside for the square was a large investment for Guangzhou which should be spent on more urgent projects, such as redeveloping roads. Secondly, the construction might adversely affect traffic and the everyday life of people around the square. Thirdly, it was proposed that in the future an underground station would be constructed under the square. It was therefore felt it would be risky to build a large-scale project before the design of the very probable station (Lin 2013). This is typical hierarchical decision-making in terms of rational calculation and risk avoidance.

From September 1998 to September 2001 the Guangzhou government spent 60.5 billion Chinese Yuan in the 'Three Changes' project (Ye 2014). By September 1999, 3,280,000 m^2 of illegal buildings had been demolished; green space had increased to 3.83 million square metres; 53.5 kms of roads had been redeveloped.

In 2003, most of the targets of three-year medium-term changes had been achieved. The building environment in some important areas, such as the two sides of main roads, had been improved. However, the communities inside and rundown areas remained unchanged; the core problems of old towns have not been overcame (Huang 2013). Mayor Lin (2013) was proud of the 'Three Changes' Redevelopment. In his book he indicates that on 18 October 2001 the Prime Minister, Mr. Rongji Zhu, had mentioned that Guangzhou was something between rural and urban space, while dirty, chaotic and poor several years ago the situation had been completely changed after the 'Three Changes' Redevelopment. Now the Pearl River in Guangzhou for him was in some sense as beautiful as the Seine in Paris (Lin 2013; p. 40).

3.2.3 Summary of the Pure Government Phase

This phase is described as the Pure Government Phase because the government had held the dominant role in urban redevelopment except for a few cases involving SOEs and redevelopment. Governance in this phase has several aspects that are similar to a hierarchical mode. Governmental actors are the main rulers in terms of managing projects; these actors are controlled by government rules and norms.

This means that minimising risks and fear of punishment are important incentives for these public actors, such as what has happened in the Garden Square Project. They prefer to make plans and design mechanisms to operate projects in a top-down process of decision-making and implementation. Expert rational consideration is important in the actors. In the planning and design activities, anticipation of obstacles and organising coordination are crucial elements. In the organisation of projects, a high-level governmental committee often plays a central position in the highest authorities and with the largest number of resources. Leaders provide commands to control subordinate actors in unilateral interaction in some cases.

However, this phase still had some characteristics similar to a market mode, such as negotiation between various actors from government, market and society was still important. Besides, actors mainly considered how to maximise their interests even when the state had already defined rules of redevelopment; competitive advantages were important in making strategies.

The government held the dominant role in the interaction between entities from the state, market and communities for a few reasons: redevelopment projects are mainly chosen by the government from political considerations; the majority of funding in redevelopment were from public resources; speed and scale of redevelopment were controlled by the government; redevelopment projects aimed to produce public goods for some specific groups, which led to less contradictions and conflicts.

3.3 Multiple Players Phase (2006–2015)

3.3.1 Overall Characteristics of Redevelopment in This Phase

In this phase, Mayor Zhang has announced that developers are welcome to enter the field of urban redevelopment in 2007 (Ye 2014). The 'Three Old Redevelopment' (TOR) policy was established in this phase which fundamentally changed the landscape of urban redevelopment through engaging more capital. Developers were once again active and contributed to the field of urban redevelopment in terms of their capital, management skills and experiences of development.

In this phase, the number of urban redevelopment projects increased; developers returned and contributed to redeveloping various urban spaces. Communities, supported by other social actors, such as mass media, NGOs and scholars, played a more important role in redevelopment. Historical preservation now had a higher priority in the governance of urban redevelopment. In 2010, the Office of the Committee to Protect the Historic City in the Guangzhou Government (历史文化名城保护委员会办公室) was established (Liu 2016). The Rules of Protecting Historical Buildings and Historical Areas in Guangzhou (广州市历史建筑和历史风貌区保护办法) and the Rules of Protecting Guangzhou as a Historic City (广州市历史文化名城保护条例) were approved in 2014 and 2015 (South Daily 2014; The Standing Committees of People's Congress in Guangzhou 2015).

The TOR policy, established in Guangzhou, Dongguan and Foshan in 2009, is the first systematic policy in municipal level to support urban redevelopment. The core of the policy is the permission to lease state-owned land through negotiation, rather than open auctions. Open auctions often lead to much higher price when negotiation has been forbidden in commercial usage land-leasing by the central government after 2004. It's a policy exception to apply negotiation in Guangzhou Province to encourage more capital to redevelop urban land.

Another crucial elements in TOR policy is the 'interests sharing mechanism' between various stakeholders, especially involving communities (Li and Liu 2018). In redevelopment of urban villages, old neighborhood and old factories, the local government set the table for developers to invest through land-leasing negotiation; community members have engaged in decision-making to share increased land income; but the renters of properties in these redeveloped sites have been excluded. The TOR policy has formulated a broader engagement between multiple players in redeveloping urban space. Particularly, in redevelopment of urban villages, communities have become the leading role within their interaction with developers and government.

The TOR policy is a ongoing process within several distinct phases; changes between these phases are dramatic (Li et al. 2022). Therefore, it is still too early analyse this continuing policy as a whole until 2022.

3.3.2 Typical Project: Enning Road Redevelopment (2006–2015)

The Enning Road redevelopment project is not a TOR project; however, it is the typical case in the third phase because it indicated the rising of community power and the formulation of multiple players in a jointed mode of governance. This project was located in the centre of an historical area in the Liwan District. On the project site buildings in the Xiguan style (the style from the west part of the historical city of Guangzhou) were everywhere. It is also the place of former homes for several famous Chinese people, such as Bruce Lee and several Guangdong Opera stars (Tan 2013). Besides these historical buildings the majority of housing in this area were private housing constructed before 1949. It is a typical old town in the city centre.

In 2006, in response to the mobilisation from the municipal governmental to improve the building environment in central Guangzhou under the spatial strategy 'adjusting the centre', Liwan District Government started the Enning Road Project (Zhou, urban planner, interview, 2013). The Enning Street redevelopment project is famous for its historical significance for Guangzhou, and its changeable process of planning and public discussion. The roles of mass media and semi-NGOs, as new elements in the field of urban redevelopment, are both crucial in such changes (Fig. 3.4).

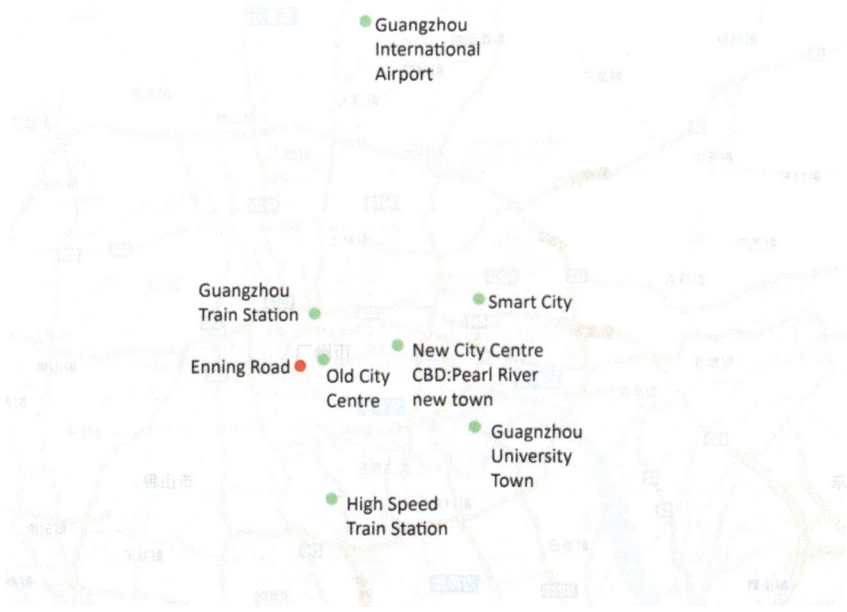

Fig. 3.4 Enning road redevelopment. *Source* author's drawing based on http://map.baidu.com/ 2016

The Enning Road project comprised 11.37 ha of land, 1,352 buildings and more than 1,900 households. Of these buildings, 297 were in public ownership, 831 were private ones, and the ownership of 224 was unclear (Wang et al. 2011). Redevelopment in Enning Road has taken place in different stages. In the first stage there were two plans in 2006 and 2007 by the district government to demolish buildings in Enning Street and store the land so cleared in its land storage. These lands could be leased to developers for fiscal income (Tan 2013; Lin, resident of Enning Road, interview, 2014).

In May 2007, the Bureau of Foreign Trade and Economic Cooperation of Guangzhou and Liwan District Government had organised the 2007 New Spring Business Conference of Guangzhou (2007 广州荔湾新春招商会) in Hong Kong. The Enning Road redevelopment project was introduced in this conference and Ruian Real Estate Ltd had expressed their interest in the project (Chen and Peng 2007).

In July 2007, the Detailed Development Control Plan (DDCP) of Duobao Road was displayed on the official website of the Bureau of Urban Planning of Guangzhou for its public participation procedure before its approval. The plan had been criticised because it would destroy the historical landscape not only by demolishing historical buildings, but also through its new planned roads with 18–26 m width to separate the whole historical area which included Enning Road.

In September 2007 an announcement about demolition was posted in Enning Street and was also met with criticism. The announcement was posted earlier than the approval of the planning document for this area and it had less consideration about historical preservation.

Under the plan in this stage, 1,950 householders would be displaced by the demolition process. The majority of the householders signed the agreement with the government agreeing to the terms of compensation, but 183 householders were unsatisfied. They wrote a public letter to the district government to protest about the planning document because there was more consideration of commercial than public interests (Tan 2013). Their motivations were questioned as it was suspected that they aimed to increase the price of compensation through the excuse of historical preservation and public interests (H2, official, interview, 2014; Y2, scholar and planner, interview, 2014).

Due to such criticism from communities and mass media, the principles of planning were modified from a focus on fiscal income to an emphasis on historical preservation in the second stage of the project. During this period, a semi-NGO (non-governmental organisation), Eninglu Concern group investigated and supported historical and cultural activities in Enning Street.

This group included university students, journalists and independent artists; they connected with external resources, such as mass media and lawyers, to help community members who insisted on staying in the partly destroyed area of Enning Street. Mass media needed news from the Enning Road Redevelopment Project to attract people's attention while the project satisfied the desire of expression to protest against the government. These householders tried to understand the purpose of mass media and used them to announce their requirements.

3.3 Multiple Players Phase (2006–2015)

In 2012, Mr. Ziliu Li, the former mayor of Guangzhou, put forward a suggestion to build a museum of traditional drama of Guangzhou. His idea was supported by the new mayor, Mr. Jianhua Chen; therefore, this redevelopment project in Enning Street seemed to become a cultural project. The government cannot obtain an income from it; but needed to spend millions of pounds on the costs of building it (Doris, semi-NGO member, interview, 2013; Xing, journalist, interview, 2013).

> Mr. Zhang as a community member knows many journalists with their mobile numbers; he really can remember these numbers. At the very beginning, he called me every time in terms of small issue such as in which family the electricity has been stopped, or the water supply has been paused. Later on, they have realised the strategy of mass media that media would not report some small issues; therefore, these residents would find some good reasons for the issues they want to be reported, not just small issues.
>
> (Xing, Journalist, interview, 12/2013).

Enning Road is significant for its involvement with social groups, such as semi-NGOs, mass media, lawyers and scholars, even those from overseas. These groups had increased the capacity of community to negotiate with the local state when developers were not participating in the main process. The local state and its sectors had to change their policy about planning, demolition and re-establishment because of pressure from residents and social forces. At the same time, a few community members made too many demands because of social support from outside.

> The function of Eninglu Concern Group in the early stage is a recorder; we record something in Enning and spread them by mass media. In the later stage, we have our own position; we often express our viewpoint to the government and mass media. Recently, because the demolishment process has been accelerated, in order to deal with these demolishment, we organise some special activities, such as setting poster in the site of demolishment with live telecast in Weibo (a Chinese social media, like Twitter). It works; the demolishing activities in this site pause.
>
> We invited some friends with lawyers background to help residents in Enning.
>
> (Doris, semi-NGO member, interview, 12/2013).

The mass media, in particular, the New Express (xinkuaibao) has played a crucial role to support communities in Enning. From the beginning, the editor of the New Express has suggested residents to write the letter to national congress to protest local government who had broken the Property Law. Besides, loads of reports have been published in newspaper, TV programs and other media to criticise the government, support community members and appeal for historical preservation.

After all the resistances against the state-led demolision and redevelopment, Liwan District has changed the plan; a new one emphasising historical protection has been approved in 2012. On the day of approval, all main newspaper in Guangzhou has put this news in their first pages. It indicated communities, with supports from media, NGOs and scholars, have become a crucial force in urban redevelopment in Guangzhou.

The new plan with more historical considerations met difficulties to attract investment due to its strict regulation and low density. After 2 years pause, the former mayor Li Ziliu has suggested to build the museum of Canton Opera in the site. Such a museum is in the process of construction after the former mayor suggested it should be built as part of the project. The redeveloped area became a chaotic place after demolition; the physical condition of the environment is horrible for the residents who insisted on remaining there. The site is still waiting for redevelopment. The Eninglu Concern Group has been disbanded because they found there was nothing to do after 2012. Community members and mass media have stayed in the site to continue their influences over redevelopment in the future.

3.3.3 Summary of the Multiple Player Phase

This phase is described as the Multiple Player Phase because the local state, the market players and communities had all played important roles with various resources and strategies. Governance in this phase has some characteristic which were similar to the network style governance. In this governing process, the relative power of entities from communities had increased their influences. It was a more balanced power structure between the state, market and communities than exhibited in the previous two phases.

The Multiple Players Phase (2006–2015) has many characteristics consistent with a network mode of cooperation. First at all, various actors from the state, market and communities coordinate their activities in terms of exchanging resources. The government is one of the players in such resource-exchange networks. Various actors have built reciprocal connections which are based on exchange but have surpassed the meaning of exchange. Trust is important in such connections and actors learn to adapt to one another's strategies and uncertain environments. Actors use persuasion to negotiate with rebellious actors in such networks, or disruptive actors are expelled if negotiation does not work. These networks have informal rules and a soft structure. Their structures can be changed, for instance, participants in the networks may change for better results and acceptance.

Therefore, the degree of flexibility in networks is medium; it is higher than in the hierarchical mode and lower than in the market mode. In these structures, actors enter into transactions with one another in multilateral ways. In these ways, actors are interdependent under the coaching and support of higher levels.

3.4 Guangzhou Style of State-Market-Community Relations

Entities from the state, market and communities have been in different relations in the three phases. These relations are reflected by social interaction between entities from the state, market and communities. These relations have formulated the Guangzhou style of state-market-communities relations. This Guangzhou style is also closely connected to the Chinese authoritarian regime.

In the Primitive Market Phase (1990–1998), the market holds the dominant role in urban redevelopment. The local state has more or less organised an informal coalition with private investors to pursue economic growth in redeveloping central Guangzhou.

In the Pure Government Phase (1998–2006), the local state is the sole player in redevelopment. Other actors from the market and communities may join these projects in different ways. However, their attendance cannot change the nature of this phase in which the Guangzhou government is the only player making major decisions and supervising other actors to achieve governmental targets. There is no coalition in this phase; market players and community members are assistants of the local state in social coordination.

In the Multiple Players Phase (2006–2015), market players have come back and entities from communities have risen up to the level of governance in urban redevelopment. These two categories of players do not just participate in the redevelopment; they also have shared power in decision-making, crucial resources and possible returns from the redevelopment. Therefore, coalitions between entities from the state, market and communities have been established to support urban redevelopment. The third phase has more characteristics as a phase in which strong coalitions operate.

References

Chen ZJ 陈志杰, Li RR 李瑞然 (2007) Guangzhou jiucheng gaizao jiantao: Qiloujie yu 2/3 laozihao xiaowang 广州旧城改造检讨:骑楼街与2/3老字号消亡 [The review of urban redevelopment in Guangzhou: Tong Lau Street and the distinction of 2/3 traditional brands]. Xinkuaibao 新快报, 21 November

Chen JT 陈锦棠 (2014) Xingtaishijiaoxia 20shijichuyilai Guangzhouzhuqu tezhengyuyanjin 形态类型视角下20世纪初以来广州住区特征与演进 [D]. [Characters And Evolutionary Process of Residential Area in Guangzhou after Early Twentieth Century: A Typo-morphological Approach]. PhD Thesis. South China University of Technology

Deng HP 邓海平 (2005) Guangzhou weifang gaizao: bingqi kaifanshang qingjie 广州危房改造:摒弃"开发商"情结 [The reconstruction of dilapidated building in Guangzhou: Abandon the "developer" complex]. Jingji ribao 经济日报, 20 May, pp. 13

Guihua 50nian-2006nian zhongguo chengshi guihua nianhui lunwenji: chenghshi fuxing 规划50年-2006n中国城市规划年会论文集: 城市复兴. Guangzhou, 21–23 September. Zhongguo chengshiguihua nianhui, Guangzhou, pp 668–673

Huang HM 黄慧明 (2013) 1949nian yilai Guangzhou jiucheng de xingtai yanbian tezheng yu jizhi yanjiu 1949年以来广州旧城的形态演变特征与机制研究 [The characteristics of urban morphological transformations and development mechanisms: a case study of Guangzhou since 1949]. PhD Thesis. South China University of Technology

Jiao HB 矫鸿博 (2010) 1979–2008nian Guangzhou zhuqu guihua fazhan yanjiu 1979–2008年广州住区规划发展研究 [A study of the development of residential district planning in Guangzhou from 1978 to 2008]. PhD Thesis. South China University of Technology

Kong WM 孔伍梅 (2008) Guangzhou jiuchengjuzhuqugaizao moshiyanjiu 广州旧城居住区改造模式研究 [The Study on Guangzhou Inner City: Residential Renewal]. Master Thesis. Sun-yat Sen University

Law of China Net (1993) *Xianggangdichan teyuezhuangao* 香港地产特约专稿 [Hongkong real estate special contribution] [online] Available at: http://www.lawofchina.net/big5/news_maz/93-9-1.html. Accessed: 22 Oct 2016

Law of China Net (1999) Yiyiergoufangkuan gaibugaitui 一億二購房款該不該退 [Should the purchase money of one hundred and twenty million be refunded] [online]. Available at: http://www.lawofchina.net/big5/news_maz/99-2-6.html. Accessed: 22 Oct 2016

Law of China Net (2001) Fupingloushi wangrideshangheng 撫平樓市往日的傷痕 [Heal the scars of the property market in the past] [online] Available at: http://www.lawofchina.net/big5/news_maz/21-4-13.html. Accessed: 22 Oct 2016

Li B, Liu CQ (2018) Emerging selective regimes in a fragmented authoritarian environment: the 'Three Old Redevelopment Policy' in Guangzhou, China from 2009 to 2014. Urban Stud 55(7):1400–1419

Li B, Yang K, Axenov KE, Zhou L, Liu H (2022) Trade-offs, adaptation and adaptive governance of urban regeneration in Guangzhou, China (2009–2019). Land 12:139

Lin SS 林树森 (2012) Renju huanjing sixiang yu Guangzhou de shijian tansuo' 人居环境思想与广州的实践探索 [Thoughts of human settlements and exploration of Guangzhou's practice]. Chengshi guihua quyu yanjiu 城市规划区域研究 2012(2):50–61

Lin SS 林树森 (2013) Guangzhou chengji 广州城记 [The stories of urban Guangzhou]. Guangdong renmin chubanshe, Guangzhou

Liu LX 刘黎霞, Fang QH 方谦华 (2010) Jiucheng gaizao zhilu 旧城改造之路 [The journey of urban renewal]. Nanfang dushibao 南方都市报, 24 November

Liu X (2006) Institutional changes for land redevelopment in transitional China: a property rights approach to the case of Jinhuajie, Guangzhou. PhD Thesis. National University of Singapore

Liu MZ 刘蠓子 (2008a) Guangzhou zuida waizi qianshuian:fangguanju beizhi cengshouyi 广州最大外资欠税案: 房管局被指曾受益 [The biggest case of owing taxes by foreign company in Guangzhou: housing authority accused had benefited]. [online] Available at: http://www.nbd.com.cn/articles/2008a-07-18/72478.html. Accessed: 22 Oct 2016

Liu MZ 刘蠓子 (2008b) Yezhu weiquan qianchu Guangzhou zuida waiqi qianshuian 业主维权牵出广州最大外企欠税案 [Right-keeping movement found the biggest case of owing taxes by foreign company in Guangzhou]. Available at: http://www.nbd.com.cn/articles/2008b-07-09/172200.html. Accessed: 22 Oct 2016

Liu Y 刘垚 (2016) Shichanghua beijingxia de chengshi gengxin 市场化背景下的城市更新 [The policy choices and roles of government in urban regeneration under the background of marketization: a case study of Guangzhou]. PhD Thesis. South China University of Technology

Pan A 潘安, Yi XF 易晓峰, Chen C 陈翀 (2006) Chengshi fanzhanzhongde jiucheng gengxin- Guangzhou de huigu he zhanwang 城市发展中的旧城更新-广州的回顾和展望 [Urban renewal in the urban development –the review and expectation of Guangzhou]

Peng X 彭昕 (2007) Guangzhou xiandai chengshi guihua fazhan jiqi fanxingde lishi yanjiu 广州现代城市规划发展及其范型的历史研究 [A historical study on development and paradigm of Guangzhou modern city planning]. M.Sc. Dissertation. Wuhan University of Technology

South Daily 南方日报 (2014) Guangzhoushichutai lishijianzhuhelishifengmaoqu baohubanfa 广州市出台"历史建筑和历史风貌区保护办法" [Guangzhou announced the rules to protect historical buildings and areas], [online] Available at: http://www.chinanews.com/cul/2014/01-16/5746226.shtml. Accessed: 30 June 2017

Tan XH 谭肖红 (2013) Chengshi guanzhi shijiaoxia de chengshigengxin juece jizhi yanjiu-yiguangzhoushi engninglu gaizao weili 城市管治视角下的城市更新决策机制研究 --以广州市恩宁路改造为例 [Research on decision making mechanism of urban regeneration from the perspective of urban governance: a case study of Enning Road in Guangzhou City]. M.A. Dissertation. Peking University

The Standing Committees of People's Congress in Guangzhou 广州人大常委会 (2015) Guangzhoushi lishiwenhuamingcheng baohutiaoli 广州市历史文化名城保护条例 [the Rules of Protecting Guangzhou as a Historic City] [online] Available at: http://www.rd.gz.cn/page.do?pa=2c9ec0233a0016bd013a00366ab30059&guid=f77f3c8211734c5fac504f68b1ec62fa&og=ff8080813f79425b013fd0f2dacd0711. Accessed: 22 Oct 2016]

References

Wang YP 王永平, Tang WJ 汤文君, Yuan DH 袁东华 (2000) Sanbian zhanlun chulun '三变'战略初论 [Discussion on the strategy of Three Changes]. Guangzhou zhengbao 广州政报 15:29–38

Wang SF et al (2011) Urban renewal planning of Enning road, Liwan District [liwanquenningluji-uchenggengxinguihua] (Guangzhou)

Wu JY 武瑾莹 (2009) Zhuguang quxian jieke hesheng xi ganggu xinjun nanfangguoji 珠光曲线借壳 合生系港股新军南方国际 [Zhuguang Corporation secretly bought a shell, Nanfang guoji became another major power of Hesengxi]. [online] Available at: http://www.guandian.cn/article/20090925/87136.html. Accessed: 22 Oct 2016

Xiang Z 橡子 (2006) Guangzhou shiwei shuji tiewan zhili loushi,zicheng zao dichanshang zenghen 广州市委书记铁腕治理楼市 自称遭地产商憎恨 [Guangzhou party secretary controls the property market and claims to be hated by the developers]. [online] Available at: http://news.sina.com.cn/c/2006-03-29/04389468195.shtml. Accessed: 22 Oct 2016

Ye HJ 叶浩军 (2014) Jiazhiguan zhuhuanxiade Guangzhou chengshi guihua-1978–2010 价值观转变下的广州城市规划1978–2010实践 [The city planning practice of Guangzhou during 1978 to 2010 based on the changing values]. PhD Thesis. South China University of Technology

Yu JT 于津涛, Sun CL 孙春龙, Huang M 黄玫 (2006) Wo bushi kaifashang de siduitou 我不是开发商的"死对头" [I'm not a rival of developer]. Liaowang dongfang zhoukan 瞭望东方周刊

Yuan LP 袁利平, Xie DX 谢涤湘 (2010) Guangzhou chengshi gengxinzhongde zijin pingheng wenti yanjiu 广州城市更新中的资金平衡问题研究 [The research of Guangzhou urban renewal funding balance]. Zhonghua jianshe 中华建设 8:45–47

Zhang YQ 张演钦 (2008) Li Ziliu: Kuojian dongfenglu wobei mazuiduo 黎子流: 扩建东风路我被骂最多 [Ziliu Li: The expansion of Dongfeng Road makes me scolded a lot]. Yangcheng wanbao 羊城晚报 6

Zhu YM 朱耀明 (2003) Jiucheng baohu gengxin de lilun yu fangfa tantao- Yi Guangzhou weili 旧城保护更新的理论与方法探讨-以广州为例 [The discussion on the theory and method of the old city protection and renewal-A case study in Guangzhou]. Zhongwai jianzhu 中外建筑 3:126–128

Open Access This chapter is licensed under the terms of the Creative Commons Attribution 4.0 International License (http://creativecommons.org/licenses/by/4.0/), which permits use, sharing, adaptation, distribution and reproduction in any medium or format, as long as you give appropriate credit to the original author(s) and the source, provide a link to the Creative Commons license and indicate if changes were made.

The images or other third party material in this chapter are included in the chapter's Creative Commons license, unless indicated otherwise in a credit line to the material. If material is not included in the chapter's Creative Commons license and your intended use is not permitted by statutory regulation or exceeds the permitted use, you will need to obtain permission directly from the copyright holder.

Chapter 4
Resilient Governance of Urban Redevelopment

Abstract This chapter aims to explain why there are diverse features associated with the different phrases. The purpose of changes in governance modes is closely related to governmental target with economic growth. Structural elements include institutional, cultural and economic ones which are external and internal challenges; changed governance modes are responses against these challenges. This chapter aim to make a clear connection between changes in the developmental environment of Guangzhou and authoritarian regime in this city. Various aspects of the developmental environment have changed in the last 25 years, and the modes of governance have also changed; however, the logic underlying the connection between these two parts has never been changed. Every mode of governance has been transformed to adapt a changed environment to maximise economic growth. This is resilient governance of urban redevelopment in Guangzhou.

Keywords Urban Redevelopment · Resilient governance · Economic growth

4.1 Changes Impacted on Urban Redevelopment in Guangzhou

Governance modes have been quite different between three phases; mayors are the key actor to determine and announce governmental strategies in every phase. Besides, other forces, such as political, economic and social changes, are also important to influence governance modes.

4.1.1 Political Changes

Structural changes have formed both constraints and resources for individuals to pursue their own interests; these constraints and resources influence behavioural patterns of governance in redeveloping Guangzhou. These changes include three categories: political factors, social and cultural elements and economic reasons. The political dimension includes crucial changes in the land administrative system, urban policy, urban planning sectors and adjustment of Guangzhou's territory.

(1) **Changes in Land Administrative System**

Land administrative agencies have responsibility to control land use, land transactions, conversion and planning issues in both urban and rural areas. Their influence on urban redevelopment is concerned with two aspects, land transactions of urban land and conversions from rural agricultural land to urban construction land.

Such transactions involve the transfer of land use rights from the local state to land users. Before 1988, land transactions were free of charge because land transaction fees were constitutionally forbidden. In April 1988, the National People's Congress approved a change in constitution to permit transaction fees, and the Land Administrative Law was modified to adopt the constitutional change. In 1990, the State Council pronounced rules to run the land transaction process with land leasing fees. In 1992, Guangzhou set up the Land Development Centre to organise land expropriation, conversions and transactions.

In land transactions, negotiation was more popular than open auctions in the 1990s. For example, in 1994 and 1996, there were 9,7405 and 10,3921 land transactions in Chinese cities, and only 12.86 and 11.51% of them were through open auctions (Zhang 2006). The use of negotiation to settle the price of land transactions means no public participation or real competition; this often leads to cheaper prices for land leasing and more profits for developers.

The Guangzhou case is part of this story; negotiation is the main method to transfer land use rights from the local state to developers. Negotiated land transactions brought about more incentives for developers to be engaged in redevelopment activities. This method also resulted in less regulation in construction because open auction often works with planning conditions. This is one of the reasons to appear the free market mode in the Primitive Market Phase (1990–1998).

On the 30th May 2001, the State Council in the Notice about Enhancing Management in State owned Land announced that every land transaction must be though open auction when there are more than two potential buyers for commercial land. Updated policy with similar rules was published in April 2003 and March 2004 by the National Land and Resources Management Ministry. After the deadline of 31st August 2004 it was no longer possible for negotiation to be used for transactions of commercial land. Therefore the 31st August, 2004 has been called 'doomsday for developers'.

Following this, land transactions became the most important resource for fiscal income in Chinese cities because land leasing fees are much higher in auctions than in negotiation (He and Wu 2005; Zhang 2006). Land leasing became a profitable business in downtown. This might be another reason for local authorities to invite developers to come back to urban redevelopment to form a de facto networking style governance in the Multiple Players Phase (2006–2015).

At the same time, conversion from agricultural land to construction land is also profitable for the local state. The local state holds a monopoly to transform collective-owned low-value agricultural land into state owned high-value construction land by expropriation with relatively low compensation, and then sell the transformed construction land to developers. However, the regulation of such conversion by the land management system became more and more strict.

From 1996 the central state started to limit the amount of land expropriation and conversion by the local state. In 1997, the central state even commanded that no expropriation could take place for the next whole year. However, local states still had methods to expropriate and convert in spite of national regulation. On the 30th July 2003, the National Council required local states to reveal their unapproved land expropriation and conversion.

In 2004 the Chinese Communist Party Central Committee announced the Number One Document about strict protection of agricultural land. In 2008, the Third Plenary Session of the 11th Central Committee approved the Decision about Some Important Problems about Supporting Reforms in Rural Areas; which claimed that the strictest protection in the world would be applied to agricultural land. Under this regulation, after 2003 the local state in Guangzhou had more difficulties in making profits from land conversion in suburban areas.

Therefore, they have paid more attention to redeveloping urban areas in the third phases of governance than in the second one, the Pure Government Phase (1998–2006). The Guangzhou Municipal Government encourages capital investment in the governmental land leasing market in terms of open auction transactions, in the Multiple Players Phase (2006–2015).

(2) **Changes as Expanding Territory After 1990**

In 2000, the territory of Guangzhou had expanded from 1,443.6 to 3,718.5 km^2. In 2005 another round of expansion had increased this number to 4,914.72 km^2. In 2014, the central government approved the increase of Guangzhou's territory to 7,434.4 km^2. The first expansion in 2000 brought about more opportunities for developing new projects in the new territory to update existing industry, because there are fewer properties in suburban areas and therefore less compensation as an obstacle to development; most of the new territory being agricultural land.

Conversion from agricultural land to construction land is problematic because of strict regulation by the central state after 2003. However, before that time, it was relatively easy to convert agricultural land. Between 2000 and 2005, expansion of urban space reduced pressure to develop old urban areas (Tan 2013). Therefore, in the Pure Government Phase (1998–2006), the local state refused private capital investors to enter the field of redevelopment in downtown areas because it was less difficult to establish new projects in the new suburban territory to update the urban economy (Fig. 4.1).

(3) **Changes in Urban Policy from the Central State**

Urban policy has influenced the scale and speed of spatial development of cities. Guangzhou is the third largest city in China, and its spatial expansion has been

Fig. 4.1 Expansion of territory from 1990. *Source* author's construct, 2023. *Notes* '1' indicates territory before 2000; '2' indicates expanded areas in 2000; '3' indicates increased urban space in 2005; '4' indicates new territory of Guangzhou in 2014

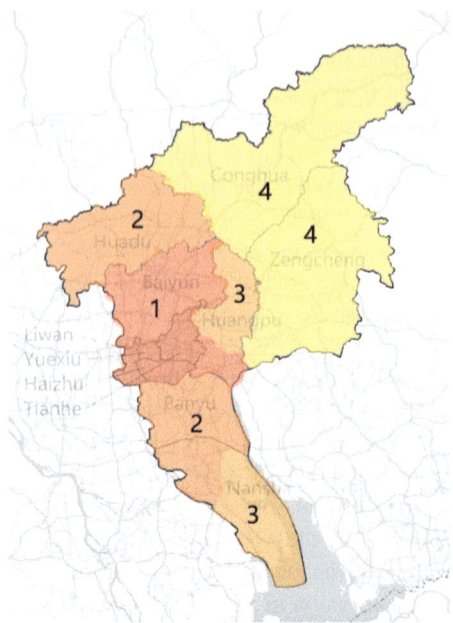

strictly controlled before 2000 due to two institutional factors. Firstly, the Socialist Five-year Plans of China, as the comprehensive planning mechanism to organise the planning economy, have a strong influence on urban issues. The sixth (1981–1985) and seventh (1986–1990) Five-year Plans announced that the sizes of large cities should be controlled, medium cities should have reasonable development and small cities need to be positively developed.

This policy has been changed in the eighth Five-year Plan (1991–1995) with the directive that sizes of large cities should be 'strictly' controlled. Secondly, the City Planning Law of the People's Republic of China has been in force from April 1990 to January 2008. This law also emphasised that sizes of large metropolis should be strictly controlled. These two institutional arrangements mean that Guangzhou as a large city had very limited opportunities to obtain new construction land; this municipality could mainly focus on redeveloping constructed land. Such institutional constraints could explain why in the Primitive Market Phase (1990–1998) urban redevelopment in downtown areas was highly active.

Such control over large cities has been released in the statement in the ninth Five-year Plan in 1996 and repeated in the City and Countryside Planning Law of the People's Republic of China in 2007. Therefore, suburban development in the next 2 phases has more advantages than in the first phase (1990–1998); redevelopment in downtown has fewer driving forces and pressures. Forbidding of developers to join urban redevelopment projects in city centre in the second phase (1998–2006) became more possible.

4.1 Changes Impacted on Urban Redevelopment in Guangzhou

When the central state aimed to control the big cities, Guangzhou has difficulties to increase its construction land.

(Yuan, scholar and planner, interview, 01/2014).

(4) Changes in Urban Planning System

The urban planning system works in a political environment; therefore, it often needs to adopt political considerations. In 2000, the territory of Guangzhou was significantly expanded. In response to this new situation of development, the local authority in Guangzhou input a new type of planning, the strategic development planning, in Guangzhou. This plan includes four parts of spatial strategy, developing to the South (南拓), optimising the North (北优), advancing to the East (东进) and combining to the West (西联).

The development to the South was the most important one because this strategy aimed to transfer Guangzhou from a domestic riverbank city to an international port city (Wu and Zhang 2007). Many mega projects, such as New University Town, Biology Island and Nansha High-tech Industrial Park had been built in the south part of Guangzhou to support this 'developing to the South' strategy; this is the spatial focus of the city in this phase; therefore, urban redevelopment in downtown areas has less importance in this period.

This changed spatial front line of economic growth might lead to the formation of the Pure Government Phase (1998–2006) because of the changed spatial focus of development. However, in 2006, Mayor Guangning Zhang announced that 'adjusting the city centre' (中调) is also critical to developing Guangzhou in a revised version of strategic development planning. It means that it is still important to redevelop the downtown area in a whole landscape of economic growth. This is the landmark of the appearance of the Multiple Players Phase (2006–2015). After that, the capital and administrative resources were distributed into declining central urban areas again. The Enning Street project had been set up for this reason (Figs. 4.2 and 4.3).

4.1.2 Social Changes

(1) **Changed Legitimacy of Governance**

There was a decline in the legitimacy of governmental behaviour after 1990. Liu (scholar, interview, 2013) claims that these residences more respected to the government than later phases. The residents trusted the real estate companies involved in this project because they were affiliated with the district government and acted as governmental agents. Therefore, in the Primitive Market Phase (1990–1998), cooperation between the state and communities was positive; conflicts mainly occurred between developers and residents.

In the third phase (2006–2015), people in the Enning Road project were skeptical of the motivation and behaviour of the government. They did not believe the purpose of this project was to improve the public good; instead, they thought the district

Fig. 4.2 Spatial strategy of Guangzhou in 2000. *Source* author's drawing based on Yuan (2015), Accessed at: 30/06/2017. *Notes* 北优 Optimising the North; 南拓 Developing the South; 东进 Advancing to the East; 西联 Combining to the West

Fig. 4.3 Spatial strategy of Guangzhou in 2006. *Source* author's drawing based on Chinese Society for Urban Studies (2010), Accessed at: 30/06/2017. *Notes* adjusting the city centre (中调) as a new strategy has been added into spatial strategy

government aimed to produce fiscal income for their own interests. Because of this change in the legitimacy of the state there were more conflicts and problems rather than coordination between government and residents. Therefore, developers become more important in the recent phase to overcome the increased distance between residents and government.

4.1 Changes Impacted on Urban Redevelopment in Guangzhou

In the Multiple Players Phase (2006–2015), legitimacy has declined and awareness of property rights has increased among individuals in communities; top-down style mobilisation of properties from individuals in redevelopment projects has become more difficult. Therefore, a de facto networking style governance becomes more necessary to organise cooperation in a less legitimated environment for the local state.

(2) **Changed Ideas About Historical Preservation Among Citizens**

In the 1990s the population in Guangzhou have less awareness about historical preservation. In the Liwan Square Project, residents were protecting economic interests, not historical issues. In the Multiple Players Phase (2006–2015), citizens, especially the young generation, have more and more awareness about historical preservation. It is a cultural demand rather than an economic consideration.

In response to such awareness mass media delivered information about the value of historical preservation. In municipal governance the local state has to positively respond to these changes in popular ideas of urban history because of strong influence from the mass media. Traditional cultural issues are supported not just by citizens, but also by the communist party and the government because history education is part of patriotism and national identity. Therefore the voices of historical preservation from the communities cannot be ignored by the local state; the interests of conserving historical buildings and areas should be included in redevelopment projects. This is also another reason to from a semi-network governance mode in this phase.

> People want to get some cultural identity at the municipal level. Some young intellectuals produced some pressure to the government in the issues (of historical preservation); however, the owners of demolished properties have fewer concerns about this.
>
> (Yuan, scholar and planner, interview, 01/2014).
>
> Because I am a local resident in Guangzhou, I have been anxious about my cultural identity when I watched news about the Enning Street project from 2008. You have a sense that the Asian Games had changed the whole landscape of Guangzhou; it was not like our memories any more. So I felt I need to pay attention to this problem and joined the Enning Focus Group.
>
> (Doris, member of semi-NGO, interview, 12/2013).

(3) **Changed Role of Mass Media**

The mass media in Guangzhou is the pioneer in terms of marketisation reform in China's mass media. Market-oriented mass media have to adjust their reports and articles to satisfy the requirements of society. After the population have changed their opinions about the legitimacy of the state and importance of historical preservation, the mass media needed to reflect such changes. Also, because Guangzhou is the capital of Guangdong Province the provincial government and municipal government have different interests in the mass media.

This is a semi check-in-balance between different levels of government; therefore, there is more freedom in Guangzhou for mass media to criticise the local state in comparison with other Chinese cities, such as Shanghai. Some mass media, such as the New Express Daily, are very active in the field of urban redevelopment. They often support the expression of community interests in redevelopment projects. These articles in the mass media are not direct challenges against the communist regime and are more tolerated compared with political protests. These functions of mass media in Guangzhou contribute to the formulation of network-style governance in the Multiple Players Phase (2006–2015), because mass media have the increased power of communities in front of the market and the state.

> Guangzhou is the capital of Guangdong; the provincial government need mass media to supervise activities of the municipal government. These media have more freedom to investigate and report. Another reason is fierce competition among media; journalists have incentives to do influential reports.
>
> (Yuan, scholar and planner, interview, 01/2014).
>
> After 2008, I have noticed that mass media has a strong influence in historical preservation issues. The television reports could produce hot topics on historical issues to the population; after that, the newspapers continue to deeply investigate stories in cultural and historical protection. Mass media makes everybody know what has happened in the field of urban redevelopment.
>
> (Doris, member of semi-NGO, interview, 12/2013).

4.1.3 Economic Changes

(1) Limited Amount of Construction Land and Increased Capital in Urban Redevelopment

At the beginning of the 1990s, capital was more important in the field of urban redevelopment than in other phases because of the scarcity of capital at that stage. This is one of the factors to explain the formulation of the Primitive Market Phase (1990–1998) where powerful market forces established and operated projects in the interaction between groups from communities and government.

In the next two decades, due to both the development of the economy and expansion of urban construction areas, there is more and more capital in the market with fewer and fewer amounts of available land for construction in urban areas. Construction land as a resource for development and redevelopment becomes relatively scarce compared with capital (see increased amount of capital in Table 4.1). Therefore, land value in urban space has increased significantly.

This increase brings about more incentives for urban redevelopment because land needs to realise its potential value. The 'three old' redevelopment policy, which is launched in the last two phases is based on this changed relative scarcity between capital and land suitable for construction. Economic prosperity is more dependent on redeveloping constructed land than converting agricultural land in terms of building office, commercial department and service industry because conversion from agricultural land to construction land has more and more strict controls.

For the government, economic growth is based on available construction land; for developers, pursuit of profits and scarcity of construction land directed their interests back to urban redevelopment. This is another reason to explain the return of private

Table 4.1 Increased capital in fixed assets in Guangzhou

Year	1990	1995	2000	2005	2010	2013
Total investment of fixed assets (construction and innovation + real estate development) in Guangzhou	9.06 billion Chinese Yuan	61.83 billion Chinese Yuan	92.37 billion Chinese Yuan	151.92 billion Chinese Yuan	326.36 billion Chinese Yuan	445.46 billion Chinese Yuan

Source Statistics Bureau of Guangzhou Municipality and Guangzhou Survey Office of National Bureau of Statistics (2014), p. 108, p. 110

capital to downtown areas in the Multiple Players Phase (2006–2015). It is also the main reason to build up the 'three old redevelopment' agenda both in Guangdong Province and Guangzhou Municipality since 2009; this agenda is an important driving force to formulate a network-style governance in the last two phases.

> Guangdong province only has 10 square kilometres increased new construction land under national land management regulation; however, the requirement for economic growth in Guangdong is for at least 40 square kilometres of new construction land. Where could we get such land? Urban redevelopment.
>
> (Deng, senior officer, interview, 12/2013).

4.2 Resilient Governance as Responses to Changes

This chapter provides explanations about the changed mode of governance in different phases. They are related to changing developmental environments. The purpose of such adaptation is to maximise economic growth under political, economic and social constraints.

In the Primitive Market Phase (1990–1998), the economic dimension, outside scale and individual aspect might be more influential than others. In this phase, political and social dimensions of this city were almost the same ones as before; they were supportive to urban growth and redevelopment. The economic difficulty, especially the scarcity of capital is the only bottleneck to urban redevelopment. To overcome this bottleneck the Mayor Li had to mobilise market force to stimulate redevelopment. The Mayor is the direct reason to develop a semi-market mode of governance. And the investment of redevelopment was mainly from outside of Guangzhou; therefore, the outside scale is more important in this phase.

In the Pure Government Phase (1998–2006), the political dimension, outside aspect might be more influential than others. In this phase, the state had much more capacities to mobilise capital in urban development; the scarcity of capital has declined. The communities had not been awaken in this period; therefore, political dimension was the most crucial in this phase to support the dominance of the local state in this phase. Mayor Lin was a powerful leader in this phase to control the governance urban redevelopment. Outside factors, such as the changed urban policy about spatial expansion, and the changes of territory in 2000 were supported his preference in governance; these factors were mainly provided by the central state.

In the Multiple Players (2006–2015), the social dimension, inside scale and structural aspect might be more influential than others. The social dimension is the most important because of the rising of communities; this rising is displayed by the active of mass media, the population's awareness of historical conservation and empowerment of communities. Such social power is mainly based on local resources rather than on outside scale forces; therefore, inside scale is more important in this period. The mayors have been less influential in this phase under the background of the rising of social forces compared with in former 2 phases. The structural changes in social interaction between the state, market and communities is more crucial than individuals, even the Mayors.

From discussion about changes in developmental environments and governance modes, it seems that governance modes can adopt changing political, economic and social environments. The spatial focus of development and relationship between the state, market and communities are reproduced incessantly in such an adoption process; different focuses of space and state-market-communities relationships are influential in governance activities.

The Primitive Market Phase (1990–1998) is focused on developing central areas of Guangzhou; development of the capital holds the dominant role in this stage. Therefore urban redevelopment is active, in which the most powerful elements are market order and market actors.

In the Pure Government Phase (1998–2006), development has been shifted from central Guangzhou to marginal areas; the government has the leading role in redevelopment. Activities in central areas are less important; public funding is not focusing on redevelopment; fewer energy appear in this field because it is controlled by the state; the hierarchical order is the mainstream order in this phase.

In the Multiple Players Phase (2006–2015), urban development pays attention to both central areas and the outskirts or outer parts of Guangzhou; communities is rising up to compete with state and market forces. The awareness of rights in the masses and the development of mass media have increased the incentives and capacities of communities to express their preferences and interests; these factors have empowered communities in negotiation in redevelopment; a more balanced power structure has developed in these two phases; network-style governance has become more important to support cooperation in urban redevelopment.

Therefore, the governance mode changed along with changes in the spatial focus of development and relationships between the state, market and communities; they interact with each other in a broad meaning of governance. Furthermore, what is the reason for these changes? The answer might be very obvious, the rationale for this adaptation is to maximise economic growth under political, economic and social constraints.

The logic of Guangzhou's authoritarian regime has been discussed in Chap. 2; however, this chapter provides more detail on urban redevelopment and its environment. Market-style governance is helpful to support economic growth in the Primitive Market Phase (1990–1998) in terms of mobilising scarce capital into redevelopment while central Guangzhou is the developmental focus; communities has less power to limit influence from market actors.

A hierarchical style of governance is necessary in a state-dominant redevelopment phase when the state focuses on suburban development and relies less on urban areas for growth; market forces have not been mobilised in this phase therefore less resources are active; hierarchical order is the most convenient mode in such an environment.

In the Multiple Players (2006–2015), economic growth is mobilised both in central and marginal areas of Guangzhou with resources from state, market and communities; the rising of communities has empowered communities; the dominance of state or market has been challenged; to mobilise more resources from various sources in an environment with more constraints requires a network-style governance to achieve growth.

4.3 Authoritarian Characteristics as the Basis for Resilient Governance

An resilient governance mode in urban redevelopment, which is pursuing maximising economic growth through adapt to changes in its developmental environment, has been displayed in 25 years as three distinct phases of governance. The reasons behind this resilient governance is the autonomy of Guangzhou to adjust its policy and actions, and its authoritarian political system to mobilise actors from the market and society, and ignoring resistance against these changes in governance.

4.3.1 Autonomies of Guangzhou Municipality

Guangzhou has more autonomy in economic activities which might lead to significant changes in governance modes in urban redevelopment when the political leaders could make their own decisions to adopt a changed environment. Guangzhou is the capital of Guangdong Province, which is the pioneer of reform after 1978 because of its geographical position far away from the central state and close to Hong Kong, the window of Western technology and institutions. Also, the leader of Guangdong Province after 1978 was Mr. Zhongyi Ren, who aimed to increase independence of Guangdong in terms of the de facto federal system between the central state and Guangdong. His requirements to build a federal system had been refused by Xiaoping Deng; however, the autonomy of Guangdong had increased in terms of decision-making in economic fields (Zheng 2013), as the capital of Guangdong, Guangzhou has relative autonomy to decide its economic issues.

This autonomy was enhanced after the decentralisation reform in 1992, the Tax Sharing System (Wu 2002). Legally, in the hierarchy of Chinese cities, Guangzhou is categorised as a large city. Cities in this category have more autonomy to impose their own laws and rules in terms of urban development, urban management and environmental protection (Chen 2014).

This relative autonomy of Guangzhou does not just lead to independent decision-making and changeable policy in governing redevelopment, but also results in specific governance modes. For example, in the 1990s, large numbers of state-sponsored enterprises, which are not all state owned, but strongly supported by agencies of the state, had built by departments of the central and local state, children of national leaders, and armies. They were located in Guangzhou and have strong connections with the municipal government.

These enterprises received resources and privileges from the local state, and in exchange, the Guangzhou government received support from the central state to enhance its autonomy. These state-sponsored enterprises, such as the Pearl River Real Estate Investment Enterprise which was built by the former mayor of Guangzhou, had more autonomy in the market, therefore, the Primitive Market Phase (1990–1996) has acted as a micro foundation to support a free market governance mode.

After 1998, because of problems of corruption in such state-sponsored enterprises, Mr. Rongji Zhu, the Prime Minister of the central state, came to Guangdong to regulate these enterprises. His activities enforcing regulation in Guangzhou might have influenced Mayor Shuseng Lin's decision to refuse developers to enter urban redevelopment in the Pure Governmental Phase (1996–2003) (Zheng 2013).

4.3.2 Mayors' Preference to Influence Urban Redevelopment Governance

Mayors, as the most influential municipal leaders in the field of urban redevelopment, have the capacity to change governance modes by their preferences. The most important features of mayors' influence in governance are flexibility and significant changes of governance modes. The diverse ages of mayors, their political experience and personalities could lead to totally different modes of governing activities.

The one party state without serious election and the dominance of the mayor in governance can explain the strong influence of mayors' preferences in governance. Furthermore, these leaders have another important significant tendency in their totally different styles in comparison with former leaders. Therefore, the Guangzhou case has two characteristics of governance.

Firstly, there are significant changes between different phases with different leaders who have the same political partisanship. The mode of governance has changed from a market mode, to a hierarchical mode, and to the mode has more networking features when the mayors have changed.

4.3 Authoritarian Characteristics as the Basis for Resilient Governance

Secondly, there are strong influences of mayors' personal preference over governance modes; these influential factors include ages, political experience and personalities. These factors are significantly formed by institutional arrangements in administrative hierarchy and the mayors' personal track in this institution.

(1) **Mayor's Age and Governance Modes**

There are rules about ages and promotion in the political hierarchy in Chinese government. On the 18th August 1980, Mr. Xiao Ping Deng, the leader of China, announced that leaders in China's communist regime need to be younger, more knowledgeable and professional. This requirement has been repeated several times after this announcement. The mayor of Guangzhou is located at the sub-provincial level in this hierarchy; to arrive at the provincial level, mayors often need to be promoted once or twice.

Mr. Ziliu Li became mayor of Guangzhou in 1990 at the age of 58. There is a '59-years-old' phenomenon in Chinese government. This means that officers who are at the age of 59 have more incentives to become corrupt because 60-years-old is the usual retirement time for ordinary officers; they prefer to grasp more economic interests when political promotion is impossible (Zhang 2009). Mayor Li's age at which he became the mayor meant that he had fewer opportunities to be promoted: this leads to more interest in economic issues rather than political considerations from a general perspective. These economic interests may lead to a less regulated style of governance in the Primitive Enterprise Phase, which could lead to more economic returns for Mr. Li.

Mr. Shusen Lin achieved the position of mayor of Guangzhou in 1996, at the age of 50. Mayor Lin had more political ambitions because he was still young to hold the position of mayor of Guangzhou. Construction in suburban areas has fewer obstacles because these areas are occupied by fewer people and a lower density of properties needs to be removed with attendant compensation issues. Therefore, in comparison with central urban areas, suburban areas have more space for mega projects and there is less resistance against such projects to support political ambitions. Redevelopment in the city centre had less priority in the governmental agenda; a pure government mode had less political risks in a less important field.

In the phase of Mr. Guangning Zhang, Mr. Jianhua Chen, as the Mayor in later periods, they have less influence compared with former Mayors.

(2) **Mayor's Political Experiences, Personalities and Governance Modes(1) Mayor's Age and Governance Modes**

Mr. Ziliu Li was appointed as the Secretary of the Chinese Communist Party Jiangmen Municipality Committee, the highest political position in Jiangmen Municipality in 1983. He has worked in Jiangmen from 1983 to 1990. After that, he was appointed as the mayor of Guangzhou Municipality. He preferred to apply his experience of relatively small cities, such as Jiangmen and Dongguan, which have more free market features in comparison with other cities. Before his phase of governance, Guangzhou had more communist characteristics and less market features because it is the capital of Guangdong and has more ideological responsibilities. After his coming to power, less control and more autonomy became the principles of Guangzhou in a Primitive Market Phase.

> You know, personal background and experience are crucial for local leaders to make decisions. Mr. Li is another example. Firstly, he has come from the position of leader in a small city, and has less knowledge to run the urban economy in the early stage of a marketing economy; therefore, he aimed to transfer the policy of small cities, such as Jiangmen, into Guangzhou with totally different scales. Secondly, he preferred to learn from Dongguan, another city much smaller than Guangzhou, which is famous for its less regulated and primitive market economy. After all, he thought Guangzhou should learn from these small cities, to improve efficiency.
>
> (Ye, senior officer, interview, 12/2013).
>
> Ziliu Li is accepted as a master in economic growth, with all the methods and support he could find; while at this phase there were less financial resources to develop. He preferred to release freedom to individuals in the market with their energy, resources and abilities. For him, the free market mode is the choice to mobilise resources to redevelop Guangzhou in such a capital-rare environment.
>
> Mr. Li is a master in economic growth. He could combine every possible element to support one priority, economic growth. At the same time, he has less methods and tools to do so compared with leaders in later phases.
>
> (Yuan, scholar and planner, interview, 01/2014).

Mr. Shusen Lin became the Deputy Secretary General of the provincial government in 1992, and later, Provincial Commission Director. The latter duty is the core department for economic growth at a macro level. His experience in these two positions helped his later efforts as a mayor. At the provincial level of government, there are more duties to balance interests between different cities, regions and sectors. Therefore, he had more comprehensive and structural considerations about urban development.

Mr. Lin preferred to be treated as an expert in the field of urban development. He wrote a book about urban development and redevelopment in 2013; and he became the honour professor in urban and regional planning department of Sun-Yat Sen University in 2014. Other mayors of Guangzhou after 1990 did not exhibit similar skills and activities. His experience in comprehensive development, and as an urban planner contributed to the formation of the Pure Governmental Phase in his period as mayor. He insisted on upholding the principle of protecting historical buildings and blocks in this phase of governance. To achieve this purpose, a pure government mode is necessary.

The logic inside municipal government had been expanded into market and communities. In this phase, redevelopment projects are chosen by the government from political considerations; the majority of funding in redevelopment were from public resources; redevelopment projects aimed to produce public goods for some specific groups, which led to less opposition and conflicts; the numbers and scale of redevelopment were very limited because of the limitation of public funding in this phase.

4.3 Authoritarian Characteristics as the Basis for Resilient Governance

> His [Mr Lin's] understanding of the metropolis is close to that of urban planners. Yes, he had the consideration as a politician; however, he very much preferred to be treated as a qualified expert about urban issues.
>
> (Ye, senior officer, interview, 12/2013).

In the phase of Mr. Guangning Zhang, Mr. Jianhua Chen, they have fewer personal references compared with former Mayors.

4.3.3 Institutional Capacities to Support Resilient Governance

This resilient governance of urban redevelopment appears for several reasons which are related to characteristics of the authoritarian regime. First of all, the local authorities have control of most of the crucial resources (He and Warren 2011), such as power in administrative approvals, setting up the orientations of urban planning and public funding to selectively support some projects, to change governance patterns. This concentration of resources on the one hand has reduced the cost of authorities to make decisions and formulate consensus; on the other hand, the concentration has strengthened the power of a few rulers to implement their decisions.

Secondly, the CCP has developed the whole system to response to external changes. These adaptive capacities include two aspects, the institutional capacities and personal abilities of the local leaders. The institutional one is related to the organised mobilisation as a Maoist heritage (Heilmann and Perry 2011). This mobilising system is operated by local leaders which are selected through the promotion system. This system is called a meritocracy mechanism by Nathan (2003). Elites are selected with related capacities in such a mechanism to control their position to operate the governance system and its changes.

Thirdly, there are fewer resistances against the tendency to change governance modes. There is no opposite party, no strong civil society and no powerful anti-growth coalition (Lai 2010), though in the Enning Road Redevelopment there is some evidence of such coalitions. The local state does not have strong obstacles against the realisation of its proposals; changing governance modes might be one of its agendas. At the same, weak barriers to change governance modes might be connected to the fragile ideology in contemporary Chinese society and CCP. Even Breslin (2011) has described China as a de-ideological society. No communist or democratic ideology can resist the tendency to change its governance modes.

The Chinese authoritarian regime has strong capacities to change its governance modes. However, why should this regime wish to do so? It is because of its dependency on economic growth to strengthen social and political stability. CCP employs economic growth and increased living standards of people to encourage political support from the population without representative rights. It is a pro-growth authoritarian governance (Lai 2010). This feature has been analysed in Chap. 2; growth is the core issue for the government in Guangzhou and other Chinese cities. This chapter has explained the reasons for changes in governance modes; keeping growth in changed environments is the reason for such changes.

After all, the Guangzhou government and other Chinese urban government have not gained their political power by fair and open elections. This means they have less legitimate power. Instead of winning support by legitimacy, they aim to increase support from society by their performance in the economic field. An authoritarian regime experiences more pressure to pursue performance. This pressure in the local state has been supported by institutional arrangements between the central and local state. The central state has provided principles and incentives to the local state; local levels have autonomies to realise the target provided by the central authority, the achievement of growth (Zheng 2013).

References

Breslin S (2011) The 'China model' and the global crisis: from Friedrich List to a Chinese mode of governance? Int Aff 87(6):1323–1343

Chen JT 陈锦棠 (2014) Xingtaishijiaoxia 20shijichuyilai Guangzhouzhuqu tezhengyuyanjin 形态类型视角下20世纪初以来广州住区特征与演进 [D]. [characters and evolutionary process of residential area in Guangzhou after early twentieth century: a typo-morphological approach]. PhD Thesis. South China University of Technology

Chinese Society for Urban Studies 中国城市科学研究会 (2010) Guangzhou baietan diqu chengshisheji 广州白鹅潭地区城市设计 [The urban design of Baietan in Guangzhou]. [online] Available at: http://www.cityup.org/case/design/20100209/59948-1.shtml. Accessed: 22 Oct 2016

He B, Warren ME (2011) Authoritarian deliberation: the deliberative turn in Chinese political development. Perspect Polit 9(2):269–289

He S, Wu F (2005) Property-led redevelopment in post-reform China: a case study of Xintiandi redevelopment project in Shanghai. J Urban Aff 27(1):1–23

Heilmann S, Perry EJ (2011) Embracing uncertainty: guerrilla policy style and adaptive governance in China. In: Heilmann S, Perry EJ (eds) Mao's invisible hand: the political foundations of adaptive governance in China. Harvard University Asia Center, Cambridge, pp 1–29

Lai HY (2010) Uneven opening of China's society, economy, and politics: pro-growth authoritarian governance and protests in China. J Contemp China 67(19):819–835

Nathan AJ (2003) Authoritarian resilience. J Democr 14(01):6–17

Statistics Bureau of Guangzhou Municipality and Guangzhou Survey Office of National Bureau of Statistics (2014) 2014guangzhou tongjinianjian 2014广州统计年鉴 [Guangzhou statistical yearbook, 2014]. China Statistic Press, Beijing

Tan XH 谭肖红 (2013) Chengshi guanzhi shijiaoxia de chengshigengxin juece jizhi yanjiu-yiguangzhoushi engninglu gaizao weili 城市管治视角下的城市更新决策机制研究 --以广州市恩宁路改造为例 [Research on decision making mechanism of urban regeneration from the perspective of urban governance: a case study of Enning Road in Guangzhou City]. M.A. Dissertation. Peking University

Wu F (2002) China's changing urban governance in the transition towards a more market-oriented economy. Urban Stud 39(7):1071–1093

Wu FL, Zhang J (2007) Planning the competitive city-region: the emergence of strategic development plan in china. Urban Aff Rev 42(5):714–740

Yuan QF 袁奇峰 (2015) Zixingche zai Guangzhou shuailuo shibushi yotade biranxing? 自行车在广州衰落是不是有他的必然性? [Is it necessary for the dying of bikes in Guangzhou?]. [online] Available at: http://www.planning.org.cn/report/view?id=133. Accessed: 22 Oct 2016

Zhang MX 张妙曦 (2006) Woguo chengshi tudi shiyongquan churang zhidu yanjiu 我国城市土地使用权出让制度研究 [A study on land transfer system in China]. PhD Thesis. Fujian Shifan University

Zhang XL 张现龙 (2009) Qianxi woguo fubai fanzui gainianzhong shuliang yinsu de yingxiang—59sui xianxiang mingti yinfa de sisuo 浅析我国腐败犯罪概念中数量因素的影响——"59岁现象"命题引发的思索 [A study of the influence of quantity in Chinese corruptions—the '59-year-old phenomenon' and its inspirations]. M.A. Dissertation. Peking University

Zheng MY 郑明远 (2003) Guangzhou ditie 1haoxian de yanxian wuye kaifa 广州地铁1号线的沿线物业开发 [The joint development long Guangzhou Metro Line 1]. Chengshi guidao jiaotong yanjiu 城市轨道交通研究 6(5):50–53

Zheng YN 郑永年 (2013) Zhongguode xingwei lianbangzhi: zhongyang- difang guanxi de biange yu dongli 中国的"行为联邦制":中央—地方关系的变革与动力 [de facto Federalism in China: reforms and dynamics of central-lical relations]. Beijing: Renmin dongfang chuban chuanmei youxiangongsi

Open Access This chapter is licensed under the terms of the Creative Commons Attribution 4.0 International License (http://creativecommons.org/licenses/by/4.0/), which permits use, sharing, adaptation, distribution and reproduction in any medium or format, as long as you give appropriate credit to the original author(s) and the source, provide a link to the Creative Commons license and indicate if changes were made.

The images or other third party material in this chapter are included in the chapter's Creative Commons license, unless indicated otherwise in a credit line to the material. If material is not included in the chapter's Creative Commons license and your intended use is not permitted by statutory regulation or exceeds the permitted use, you will need to obtain permission directly from the copyright holder.

Chapter 5
Conclusion

Abstract This chapter aims to conclude the whole study. This research employs the case of Guangzhou to explore the forces governing urban redevelopment in Chinese cities which have experienced rapid development and radical transformations in the last few decades. Based on studies of institutional background as authoritarian regime, modes of governance in redeveloping Guangzhou, and connection between authoritarian political system and three distinct governance phases, a resilient governance in redeveloping Guangzhou has been preliminarily formed. This research has difficulties in terms of a structural narrative about stories in Guangzhou over 25 years; it is difficult to collect data and make choices among data. At the same time, Guangzhou is just one case among Chinese cities; it is acknowledged that it is difficult to analyse the nature of governance in Chinese cities by the example of one city. To overcome these research limits and develop this research, studies in the future need to follow the latest developments, broaden the research field and deepen the nature of the research.

Keywords Urban Redevelopment · Institutional background · Governance mode · Resilient governance

5.1 Research Findings

In this research, there is a clear connection between governance in redeveloping Guangzhou and Chinese style authoritarian regime. This connection is the key to understand resilient governance in Guangzhou. And it is the focus of this research.

5.1.1 Institutional Background

The institutional background governing urban redevelopment is based on the governmental relationship between Guangzhou and Beijing. At the organisational level, the 1994 reform of distribution of fiscal income between the central and local state resulted in fewer resources at the municipal level with more duties, such as requirements to provide a healthy lifestyle, education and urban infrastructures. The Guangzhou government needed to increase its fiscal income to fix the mismatch between fewer resources and more duties.

At the individual level, the municipal leaders were eager to be promoted by the leaders in the central level; this promotion depended on local economic performance to some degree. Therefore, the local state had strong incentives to create a pro-growth regime in Guangzhou. These incentives are supported by a group of institutional arrangements.

Housing reform after 1998 stopped the free distribution process of housing in work-units; people had to buy commercial properties in the real estate market. This market became prosperous after 2000 as indicated by the significantly increased property prices. This prosperity led to enormous demand in the land-leasing market where the local state was the only seller holding a monopoly on rents. Income from the land-leasing market contributed to fiscal income and supported economic growth at the municipal level.

Urban redevelopment plays a crucial part in both the real estate market and land market. Urban planning, as a professional tool employed by the local state or developers, could increase land values in terms of high-quality design, advertisements, high density and changing land functions. In such an institutional framework, a land-oriented growth and redevelopment context has been formulated.

5.1.2 Three Phases of Governance in Urban Redevelopment

Based on the institutional background, governance modes can be analysed with empirical data from Guangzhou. Attention here focuses on what has happened in urban redevelopment. There are three phases with different patterns of governance: 1 the Primitive Market Phase (1990–1998); 2 the Pure Government Phase (1998–2006); and 3 the Multiple Players Phase (2006–2015).

The Primitive Market Mode (1990–1998) was close to the market mode of governance in several ways: a huge amount of small-scale developers were active in redevelopment with less exclusive control or monopoly in the hands of a single entity; the redevelopment process had relaxed regulations in terms of planning; the use of a contract was the main method to define relationships between entities; negotiation was the main mechanism to solve conflicts; developers had relatively more power in terms of decision-making in different stages.

The Pure Government Phase (1998–2006) had a semi-hierarchical mode of governance. Agendas of redevelopment were established by the local state mainly by political considerations; the majority of investment in projects was supported by public funding; construction works were contracted out to construction companies under the supervision of governmental sectors; a fewer number and smaller scale of redevelopment projects were built in this phase because of the limited amount of public investment.

The Multiple Players Phase (2006–2015) have more characteristics close to the network mode of governance. This Phase involved more diverse players in the process of redevelopment, such as actors from mass media and voluntary sectors. Their attendance had increased resources of social groups from redeveloped communities; exchanges of resources between public and private sectors were more frequent. Therefore, entities from the local state, market and communities are interdependent in their interaction; various coalitions between groups were established. However, the local political leaders were still powerful exercising control of the direction of redevelopment even when other actors had become stronger than before.

5.1.3 Resilient Governance of Urban Redevelopment in Guangzhou

The third part of research aims to provide understanding of the purpose of changes in governance mode; attention focuses on why such changes have happened. The main purpose here is to pursue economic growth in various developmental environments. First of all, Guangzhou is the capital of the Guangdong Province and the pioneer of reform after 1978; these factors bring about more autonomy in the Guangzhou government to make their own decisions in governing redevelopment. Therefore the governance mode in Guangzhou has more possibilities for change.

In the 1990s, the local state had less regulation than later to control the land-leasing process. In such processes, negotiation was the most popular method to transfer land use rights to developers. This method often meant cheaper prices for land and more profitable prospects for redevelopment. This factor contributed to the formulation of the semi-market mode of governance in the Primitive Market Phase (1990–1998) when market forces were active and dominant in this phase.

At the same time, the lack of capital in Guangzhou also supported the dominant role of market forces in this phase. National urban policy at this stage had restricted the extension of urban territory, which led to more incentives to redevelop old towns than to develop new cities. After 2000, the territory of Guangzhou was extended from 1443.6 to 3718.5 km^2. New space for development had become available while the cost in developing rural areas was cheaper than redeveloping old urban areas because there were fewer compensation fees in rural space.

The urban planning system in this stage had provided guidance to urban sprawl towards the south and east of Guangzhou. Therefore, urban redevelopment in old towns had less significance for construction and growth; the local state had dislodged developers from urban redevelopment and redevelopment had become less active in the semi-hierarchical mode of the Pure Government Phase (1998–2006).

In the next phase, the Multiple Players Phase (2006–2015), stricter and stricter control over conversion from agricultural land to construction land in rural areas meant redeveloping old towns was more reasonable than before. Urban redevelopment became a hot spot again. In these two phases network style governance has involved more social actors such as communities, voluntary organisations and mass media. This is due to the rise of market style mass media and the awareness of historical preservation in Guangzhou after 2003.

The popular idea of legitimacy of governmental activities has also changed. The general population have become suspicious of the behaviour of the local state; the municipal government needs to empower the communities to express their interests in legitimate governmental behaviour.

In addition to these structural factors, the mayors of Guangzhou have strong influences on governance modes; especially the mayors in the first two phases, Mr. Ziliu Li and Mr. Shusen Lin, who have significant styles of governance. Their ages, political experiences, personalities and preferences have strongly contributed to the governance styles adopted in the first two phases. The influence of the mayors is characteristic of the Guangzhou case.

Various factors have different importance in diverse phases. In the Primitive Market Phase (1990–1998), factors from the economic dimension, outside scale and individual aspect (mayors) might be more influential than others. In the Pure Government Phase (1998–2006), the political dimension, outside scale and individual aspect might be more influential than others. In the Multiple Players (2006–2015), the social dimension, inside scale and structural aspect might be more influential than others. The relative importance of these factors has been discussed in '6.6 conclusion: adaptive governance mode for growth'.

Every factor mentioned above has influenced the governance modes under the pro-growth mechanism. This means the governance mode in specific phases has been changed to adapt to changing political, economic and social circumstances. The purpose or logic for this adaptation is to maximise economic growth under political, economic and social constraints. The modes of governance have changed dramatically while the logic behind these modes has not changed. However, this section has only explained the incentives and directions of changes in governance modes; the possibilities and capacities of Guangzhou to change governance mode to fit changing circumstances needs further investigation.

5.2 Comparing Guangzhou to Other Chinese Cities

The Guangzhou case is part of the transformation of Chinese cities in the last 25 years. Table 5.1 above summarises the process of urban redevelopment in China which seems to be a gradual transformation from a planning system to a market system, from a centralised system to a more decentralised system, with more and more factors and social groups included in such a process Wu (2015).

5.2 Comparing Guangzhou to Other Chinese Cities

Table 5.1 Periodization of Chinese urban redevelopment

Periodization	Historical conditions	Policy aims[a]	Approaches and dominant actors	Products	Impacts
1979–1997	Economic reform in 1979	Dilapidated housing renovation	Capital mobilization by state work-units Subsidized housing purchase by employers	Local government funded housing renewal Relocated housing estates in suburbs	Sporadic upgrading Modest improvement
1998–2008	Asian financial crisis in 1997 WTO membership in 2001	Housing renewal + Land revenue generation	Property developers Entrepreneurial local states	Commodity housing estates and condominium Suburban new towns	Residential Relocation Suburbanization Displacement of inner urban residents
2009-present	Global financial crisis in 2008	Urban spatial order Land revenue generation + Economic restructuring	Property developers Entrepreneurial local states State-backed investment platforms and corporations	High-tech parks New mix use complex (office, shopping, for example, shopping malls) Large-scale superblocks Master-planned estates	Suburbanization Displacement of inner urban residents Urban restructuring Displacement of migrant tenants

Note [a]For new policy aims, these are noted with "+." WTO = World Trade Organization
Source author's construct, 2016. Wu (2015), p. 9

In Wu's theory, the years after 1990 can be divided into three phases. The first one is before the 1998 Asian Financial Crisis. In this phase, capital from work-units and public funding have improved the living conditions of citizens on a limited scale. There is no consideration of income from the perspective of the local state.

The second phase (1998–2008) has involved the local authority in the pursuit of fiscal income by land-leasing. The entrepreneurial local state has been formulated. Developers have become active in this phase.

In the third phase after 2008, the local state focuses more concern on the upgrading of the local economic structure in redevelopment in addition to pursuit of direct fiscal income. State backed investment platforms have been established as important financial approaches to support local growth.

In another paper, Zhai and Wu (2009) have summarised urban redevelopment in Chinese cities from 1949. They maintain that the phase after 1990 should be divided into two phases, a transition phase with real estate development (1990–2000), and rapid and comprehensive redevelopment after 2000. In the first phase, real estate development has become an important strategy for the local authority to stimulate economic growth. The local state functions as an economic actor rather than a political actor. Institutions involved in redevelopment have been improved. More and more stakeholders have been involved. In the second phase (2000 to date) there are more comprehensive and sustainable elements in urban redevelopment, such as a harmonious society and sustainable cities which are important targets.

These two divisions of the phase have some similarities because both of them tried to connect the urban redevelopment of Chinese cities with national and global changes to adapt to such changes, especially the global elements, such as the financial crisis in Asian counties. In these statements about the redevelop Chinese cities, the most important similarity is the pursuit of economic growth by the local authorities. They are entrepreneurial actors aiming to maximise their interest in governing urban activities. Secondly, developers in the real estate industry are important actors in growth-oriented redevelopment projects. Cooperation between the local state and developers is a common phenomenon. Thirdly, more and more elements are included in urban redevelopment, such as financial departments and social entities.

The Guangzhou case has some similarities with the ideas of Zhai and Wu (2009) and Wu (2015). Firstly, the state plays a central role in any period. The central state provides the national policy and builds up a context for local development; the local state employs different strategies to adapt to changed environments, nationally or globally. Secondly, the tendency of changing governance modes leads towards a more comprehensive and inclusive redevelopment process. More social and cultural elements are included in governance practice. Thirdly, such a tendency for changes appears more at the level of governance, but not at the level of meta governance. The logic and dynamics of designing, context and reproduction have not been changed. For instance, Wu's (2015) statement about the local changes follows the logic of the entrepreneurial local state in any phase of redevelopment.

Compared with other Chinese cities, Guangzhou has two main differences in terms of governance. Firstly, it is more fragmented than other Chinese cities in terms of the distribution of resources and power. For instance, Guangzhou has more urban villages than other Chinese cities, especially compared with Shanghai. This is because the Guangzhou government is more tolerant than other local authorities; more urban villages have been demolished in other Chinese cities.

Existing urban villages in Guangzhou have their own collectively owned land and massive properties in these lands. These lands and properties have large economic values within the villagers' hands. Such values are fragmented into villages and bring bargaining power to the villages when they need to negotiate with local authorities. For this reason, demolishing these villages is much more difficult than doing so in Beijing and Shanghai. Therefore, villages have joined the coalition of redevelopment. The Guangzhou governance is more inclusive for non-public actors.

Secondly, the Guangzhou case displays a more discontinued picture about governing urban redevelopment between phases. Other Chinese cities have more continuous patterns of governance. This is because the Guangzhou governance style is more resilient than in other Chinese cities. Behind the market, hierarchy and network mode, there is a market mode as meta-mode to change its governance mode between various modes in specific contexts. This market-oriented meta-mode has several aspects.

Guangzhou is more adaptive to the changed environment by changing its governance modes. This is market-style behaviour in a real market environment while hierarchy and network have less flexibility in changed circumstances. In Guangzhou market entities have always been active, even in the Pure Government Phase (1998–2006). In this phase, the local authority operates urban redevelopment by market-style governing skills in mega project such as the Inner Loop Redevelopment.

In addition, private developers have found opportunities in this phase. They operate redevelopment in SOEs which are not controlled by the local authorities. Some of these developers, such as Fuli, have grown up into the largest real estate enterprises in China. Fili won its first capital in redeveloping old SOEs in the Pure Government Phase.

5.3 Limits of This Research and Areas for Future Research

This research displays the political nature of urban redevelopment in Guangzhou from 1990 to 2015 from the perspective of governance. There are two major limits of my research. First, it is a large scale research which includes hundreds of square kilometres of space and spans a time of 25 years. It is very challenging to analyse such a temporal and spatial scale in one research; so many details about actors, institutions, events and projects have happened. This huge amount of information is hard to collect; after collection, choosing among the collected data is also difficult; displaying the chosen data to form a structural narrative is full of uncertainty. There might be a large amount of data which have been missed because of the limits of time, resources and capacity of the researcher.

Secondly, governing urban redevelopment in Guangzhou is one case of urban governance in Chinese cities in the last 25 years. Do other Chinese cities have similar processes and mechanisms or not? Because my research is focused on Guangzhou and has less concern with other cities, even if there is a comparison between Guangzhou and other Chinese cities this comparison is unbalanced, much more attention has been directed to Guangzhou than to other cities. The answer to this question is unclear. The original aim of this research is to investigate the nature of urban politics in the Chinese authoritarian regime; the answer from urban redevelopment might be problematic.

With some limits, the whole research displays that the governance mode of Guangzhou has been changed several times to adapt to changed political, economic and social environments. This mode is resilient to maximise growth in various conditions. This resilience seems to increase the capacity of the communist regime to continue its ruling by economic performance.

In the changing process of governance, it is more and more open, inclusive and transparent; such changes are progressive according to Western standards. However, such a strong capacity of adaptation and the process of improvement might meet their limitations. It means changes in the future may have conflicts with the essence of the one-party state system. Therefore, 'where is the limit of adaptive capacity of China's authoritarian regime' is an interesting question. Both prosperity and the collapse of this country are possible.

This question requires further research; it also needs more time to observe what the future for such a special regime is. It is also a great experience for a human being to explore the possibility of governance. This is the first possible direction of future research about the latest development in governing urban redevelopment.

At the same time, it might be interesting to expand this research to evaluate results of urban redevelopment. This research is more about the process of redeveloping; however, the cause-effect relations between processes and consequences are crucial to understand the nature of governance in redevelopment. Evaluating economic, social, cultural and environmental outcomes of redevelopment projects are important to improve the governance activities as practical issues. This is the second possibility for further study in terms of expanding research to other topics in the same field.

Furthermore, this research is more about power relations in interaction in urban redevelopment among entities from the state, the market and communities; however, the production of knowledge in these redevelopments is also important. This is because knowledge is essential for actors to make decisions in the governance structure; it is a deeper level of governance. This is the third potential research agenda in terms of understanding the deeper nature of governance. In short, the future research should follow the latest developments in reality, expand the field of research to unknown areas and look in more depth at the nature of governance. They are all middle-scale research objectives.

If I can go further, the next crucial questions will be: Is it possible to surpass the dualistic discussion between authoritarian and democratic governance in human society to discuss appropriate governance modes? If it is possible, what are the fundamental elements to answer such a question? In fact, this book aims to explore these two questions; but I do not have the capacity to find appropriate answers. It will be interesting to see what can be found in the future.

Reference

Wu F (2015) State dominance in urban redevelopment: beyond gentrification in urban China. Urban Aff Rev 27:1–28

Open Access This chapter is licensed under the terms of the Creative Commons Attribution 4.0 International License (http://creativecommons.org/licenses/by/4.0/), which permits use, sharing, adaptation, distribution and reproduction in any medium or format, as long as you give appropriate credit to the original author(s) and the source, provide a link to the Creative Commons license and indicate if changes were made.

The images or other third party material in this chapter are included in the chapter's Creative Commons license, unless indicated otherwise in a credit line to the material. If material is not included in the chapter's Creative Commons license and your intended use is not permitted by statutory regulation or exceeds the permitted use, you will need to obtain permission directly from the copyright holder.

If you have any concerns about our products,
you can contact us on
ProductSafety@springernature.com

In case Publisher is established outside the EU,
the EU authorized representative is:
**Springer Nature Customer Service Center GmbH
Europaplatz 3, 69115 Heidelberg, Germany**

Printed by Libri Plureos GmbH
in Hamburg, Germany